THE ULTIMATE BOOK OF
INCREDIBLE FACTS

1000 FASCINATING FACTS AND QUIZZES ABOUT SPACE, ANIMALS, MYSTERIES, POP CULTURE, AND EVERYTHING IN BETWEEN

Merry Christmas Kathross & Sophia!

J Kimberly

For Saturn, my loyal companion and endless source of joy. Your curiosity and boundless energy inspire me every day to explore the wonders of the world.

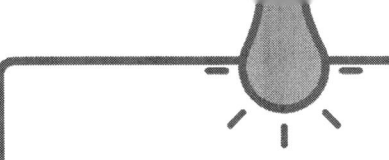

TABLE OF CONTENTS

INTRODUCTION

Welcome to an exciting and unforgettable journey through 1,000 of the most fascinating, funny, mind-bending, and downright bizarre facts you'll ever read! This book isn't just a simple collection of trivia—it's your ultimate guide to uncovering the strange, the spectacular, and the downright unexpected curiosities that make our world so intriguing. Whether you're a trivia enthusiast, a curious mind always seeking new knowledge, or just someone looking for quirky fun facts to impress your friends and family, you're in for a wild and thrilling ride!

From the wonders of nature to the mysteries of space, from prehistoric giants to future technologies, and from hilarious pop culture moments to eerie ghostly encounters, this book has something for everyone. You'll laugh, you'll learn, and you might even question reality a few times. Did a chicken really live for 18 months without its head? Can dinosaurs still surprise us millions of years later? And will AI eventually become our funniest stand-up comedians? The answers: yes, absolutely, and let's just hope they don't start charging for tickets.

Each chapter dives into a unique theme, packing in 100 incredible facts and finishing with a pop quiz to test what you've learned. Whether you're flipping through at random or tackling the book chapter by chapter, get ready for jaw-dropping revelations and plenty of "wait, what?!" moments.

So, grab your favorite snack, find a comfy spot, and prepare to have your mind blown, one fact at a time. Let the adventure begin!

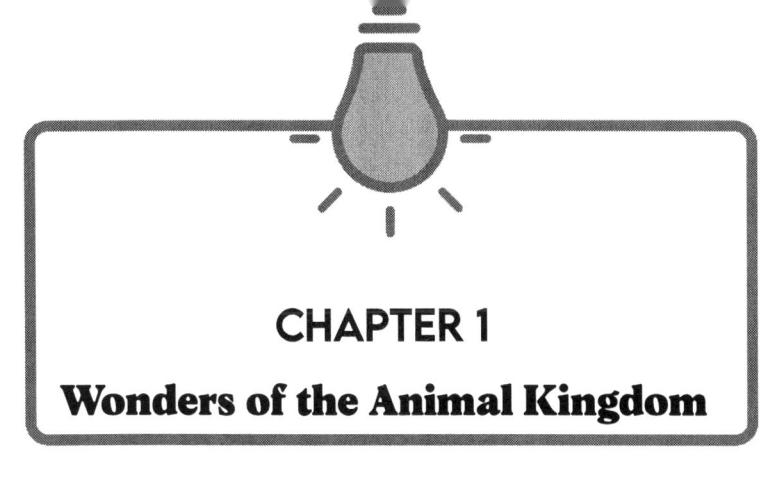

CHAPTER 1

Wonders of the Animal Kingdom

Animals are nature's ultimate comedians, survivalists, and daredevils all rolled into one. From bizarre survival strategies to hilarious quirks and mind-blowing abilities, the animal kingdom never fails to amaze. Whether they're navigating unique challenges or showing off their peculiar talents, this chapter dives into the fascinating and sometimes downright unbelievable lives of our planet's most remarkable creatures.

Fact #1: Sloths Can Mistake Their Own Arms for Branches

Sloths are so slow and lazy that they sometimes grab their own arm, thinking it's a tree branch, and then accidentally fall out of the tree. Talk about an *oops* moment!

Fact #2: Wombat Poop is Cube-Shaped

Yes, wombats poop cubes. Their unique digestive system compacts their waste into little squares, so it doesn't roll away. They even use it to mark their territory. Nature's most creative bathroom strategy!

Fact #3: Male Giraffes Drink Pee to Find a Mate

To figure out if a female giraffe is ready to mate, males take a mouthful of her pee. It's gross, but apparently, it gets the job done. Who needs flowers and chocolates when you have giraffe logic?

Fact #4: Octopuses Have Three Hearts, and Two of Them Stop When They Swim

Octopuses are so chill that two of their hearts stop beating when they swim. It's like their bodies say, "Nah, we don't need to work that hard." True masters of taking it easy.

Fact #5: Dolphins Call Each Other by Name

Bottlenose dolphins have unique whistles that act like names. They call each other with their "signature whistles," proving they're not just cute but also better at remembering names than most of us.

Fact #6: Parrots Name Their Babies

Speaking of names, parrots are also pros at personalizing things. Mama parrots give their chicks unique calls, which act as baby names. Nature's version of yelling, "Come here, Timmy!"

Fact #7: Sea Cucumbers Can Puke Their Guts Out as a Defense

When attacked, sea cucumbers literally throw up thoir guts to distract predators and make a quick escape. They eventually grow everything back, but talk about a dramatic exit!

Fact #8: Penguins Propose With Rocks

Male penguins give their chosen mate a pebble as a proposal gift. If she accepts, they use the pebble to build their nest. Basically, it's the penguin version of a diamond ring.

Fact #9: Butterflies Taste With Their Feet

Butterflies don't need forks or spoons—they taste everything with their feet. Imagine stepping on pizza and instantly knowing if it's pepperoni or cheese. Handy… or footy?

Fact #10: Pistol Shrimp Have a Sonic Boom Claw

Pistol shrimp snap their claw so fast it creates a bubble hotter than the surface of the Sun and louder than a gunshot. These tiny creatures are packing serious heat—literally.

Fact #11: Male Koalas Sound Like Demons During Mating Season

Koalas may look adorable, but during mating season, males make a deep, growling noise that sounds more like a haunted house than a cuddly bear. First impressions can be deceiving!

Fact #12: Axolotls Can Regrow Their Brain

The axolotl, also called the "smiling salamander," can regrow pretty much any part of its body—even parts of its brain. Forget nine lives—these guys just regenerate!

Fact #13: Kangaroos Use Their Tail as a Fifth Leg

Kangaroos don't just hop around aimlessly. They use their tails as an extra leg when they walk, making them nature's version of a pogo stick.

Fact #14: Puffins Have Glow-in-the-Dark Beaks

Under UV light, puffins' beaks glow like neon signs. Turns out, they're not just cute—they're also ready for a rave.

Fact #15: Elephants Can Use Their Trunks Like Straws

Elephants can suck water up their trunks—up to two gallons at a time—and then squirt it into their mouths. It's basically the biggest straw on the planet.

Fact #16: Cows Have Best Friends

Cows are social animals and often form strong friendships. When separated, they show signs of stress. Basically, they're just big, lovable BFFs.

Fact #17: Narwhals Use Their "Unicorn Horn" as a Toothbrush

The narwhal's tusk is actually a giant tooth. Scientists believe they use it to clean off parasites and show dominance. Who knew brushing teeth could be this hardcore?

Fact #18: Oysters Can Change Gender

Oysters switch genders depending on what helps them reproduce more efficiently. Nature's motto: flexibility is key.

Fact #19: Wood Frogs Freeze and Come Back to Life

Wood frogs in Alaska freeze solid during winter, even stopping their hearts. When spring arrives, they thaw out and hop away like nothing happened. Nature's ultimate reboot.

Fact #20: Hippos Sweat Sunscreen

Hippos secrete a red, oily substance that acts as both sunscreen and an antiseptic. Turns out, they've been doing skincare better than us all along.

Fact #21: Goats Have Accents

Goats don't just bleat—they bleat with style. Depending on their social group, goats develop unique accents, making their voices as individual as human ones.

Fact #22: Tasmanian Devils Yawn When They're Nervous

Tasmanian devils show off their ferocious teeth by yawning when they're scared or anxious. It's the animal version of a nervous smile.

Fact #23: Starfish Can Vomit Their Stomachs to Eat

Starfish take "eating out" to the next level—they push their stomachs out of their bodies to digest prey, then pull them back in. Bon appétit?

Fact #24: Pigeons Can Do Math

Pigeons might poop on statues, but don't underestimate them—they're surprisingly good at basic math, solving problems as well as some primates.

Fact #25: Flamingos Bend Their Legs Backward

Flamingos' knees aren't where you think. What looks like a backward knee is actually their ankle! Their real knees are hidden closer to their bodies.

Fact #26: Octopuses Punch Fish

Octopuses sometimes punch fish out of pure spite during hunting trips. Apparently, teamwork doesn't come naturally to these sea creatures.

Fact #27: Bees Communicate Through Dancing

Bees perform a "waggle dance" to share directions to food. Imagine explaining directions by busting a move—nature's original GPS!

Fact #28: Frogs Close Their Eyes to Swallow

Frogs use their eyes to help push food down their throats. They literally close their eyes while eating—talk about using everything you've got!

Fact #29: Snakes Can Hibernate Together in Giant Balls

To stay warm in winter, snakes hibernate in massive groups called "hibernacula," forming writhing, slithery snake balls. Not creepy at all, right?

Fact #30: Ants Stretch When They Wake Up

Ants have morning routines, too. When they wake up, they stretch their tiny bodies to get ready for the day. Even bugs need a good stretch!

Fact #31: A Shrimp's Heart is in Its Head

Shrimps wear their hearts on their sleeves—or in their heads, to be exact. Their heart and brain are neighbors in their tiny crustacean bodies.

Fact #32: Sea Otters Hold Hands When They Sleep

Sea otters hold hands to keep from drifting apart while they snooze. It's basically the most adorable thing in the animal kingdom.

Fact #33: Camels Have Three Sets of Eyelashes

Camels sport three layers of eyelashes to protect their eyes from sandstorms. Take that, mascara commercials!

Fact #34: Koalas Nap for 22 Hours a Day

Koalas are basically the ultimate couch potatoes, spending up to 22 hours a day sleeping. The remaining time is dedicated to eating and looking cute.

Fact #35: Crows Can Hold Grudges

Crows are super smart and will remember people who annoy them. Even worse, they'll tell other crows about you. Watch your back!

Fact #36: Chickens Are Masters of Math

Chickens can count and even perform basic arithmetic. Who knew the phrase "chicken scratch" could apply to math, too?

Fact #37: Goats Love to Climb Trees

In Morocco, goats climb trees to feast on argan fruit. Imagine looking up and seeing a tree full of goats—it's as weird as it sounds!

Fact #38: Pandas Do Handstands to Pee Higher

Male pandas mark territory by peeing as high as possible on trees. To achieve maximum height, they perform handstands while they go. Bravo, pandas!

Fact #39: Sharks Have Been Around Longer Than Trees

Sharks existed over 400 million years ago, even before trees evolved. Basically, they're ancient swimming legends.

Fact #40: Turtles Breathe Through Their Butts

Some turtles, like the Fitzroy River turtle, can absorb oxygen through their butts. Nature really said, "Why not?"

Fact #41: Axolotls Always Smile

Axolotls, the adorable amphibians that can regrow limbs, always look like they're smiling. It's their secret to being everyone's favorite pet.

Fact #42: Caterpillars Melt Into Goo Before Becoming Butterflies

During metamorphosis, caterpillars turn into literal goo inside their cocoons before transforming into butterflies. Talk about a glow-up.

Fact #43: Squirrels Forget Where They Bury Their Nuts

Squirrels forget about 50% of the nuts they bury, accidentally planting trees everywhere. Thanks for the forests, squirrels!

Fact #44: Ducks Sleep With One Eye Open

Ducks can shut down half their brain and keep one eye open while sleeping to watch for predators. Nature's ultimate multi-taskers.

Fact #45: Some Ants Explode to Defend Their Colonies

Malaysian exploding ants will literally blow themselves up to protect their colonies, releasing sticky goo that traps enemies. Ants go hard.

Fact #46: Jellyfish Can Survive Without Brains

Jellyfish may not have brains, but they've been thriving for millions of years. Proof that you don't need brains to succeed—just go with the flow.

Fact #47: Frogs Can Throw Up Their Entire Stomachs

When frogs need to get rid of something nasty they ate, they vomit their entire stomach, clean it out with their hands, and pop it back in. Talk about extreme tidying up!

Fact #48: Pigeons Can Recognize Themselves in Mirrors

Pigeons are one of the few animals that can recognize their own reflection. So next time you see a pigeon staring in a window, it might just be admiring itself.

Fact #49: Crabs Can Grow Back Their Legs

Crabs don't sweat losing a limb—they just grow it back. Imagine that happening after stubbing your toe!

Fact #50: Snails Can Hibernate for Three Years

If conditions aren't ideal, snails can take the ultimate nap and hibernate for up to three years. Now that's commitment to rest.

Fact #51: Platypuses Don't Have Stomachs

Platypuses skipped the whole stomach thing—they digest food directly in their intestines. Because who needs an extra step?

Fact #52: Dogs Have "Fingerprint" Noses

Every dog's nose print is unique, just like human fingerprints. This means your pup could technically solve crimes!

Fact #53: Crocodiles Can't Stick Out Their Tongues

Crocodiles' tongues are attached to the roof of their mouths to stop them from biting themselves during their powerful chomps. Smart move.

Fact #54: Goldfish Have Better Memories Than You Think

Goldfish can remember things for months, not seconds as the myth suggests. So yes, they remember when you're late feeding them.

Fact #55: Moose Can Dive Underwater

Moose are excellent swimmers and can dive up to 20 feet underwater to feed on aquatic plants. Bet you didn't expect that!

Fact #56: Bees Can Detect Explosives

Bees are so good at smelling things that they can be trained to detect explosives. These tiny insects are basically nature's bomb squad.

Fact #57: Tarantulas Can Keep Tiny Frogs as Pets

Some tarantulas keep frogs around to eat pests that could harm their eggs. It's like having a personal pest control service.

Fact #58: Male Seahorses Give Birth

In a role-reversal, male seahorses carry and give birth to the babies. A true "dad of the year" award goes to them.

Fact #59: Owls Can Twist Their Heads Almost All the Way Around

Owls can rotate their heads 270 degrees without breaking a sweat. It's the ultimate party trick.

Fact #60: Some Birds Sleep While Flying

Albatrosses can take short naps mid-flight, staying airborne for months at a time. Multitasking at its finest.

Fact #61: Lobsters Pee Out of Their Faces

Lobsters have urinary openings right under their eyes and use pee to communicate during fights. Nature works in mysterious (and gross) ways.

Fact #62: Koalas Have Fingerprints Like Humans

Koalas' fingerprints are so similar to humans' that they could confuse crime scene investigators. Don't worry, they have solid alibis.

Fact #63: A Jellyfish Can Live Forever

The Turritopsis dohrnii, known as the "immortal jellyfish," can revert to its juvenile form after reaching maturity, essentially restarting its life cycle. It's nature's version of hitting the reset button.

Fact #64: Hedgehogs Cover Themselves in Spit

Hedgehogs chew on things they like, then spit all over themselves. Scientists aren't sure why, but maybe it's their version of cologne.

Fact #65: A Group of Flamingos is Called a "Flamboyance"

It's only fitting that these bright, stylish birds have a name as fabulous as their looks.

Fact #66: Cows Moo in Regional Accents

Depending on where they're raised, cows develop unique moos that match their environment. A Southern cow might moo with a drawl!

Fact #67: Birds Can Sleep With Half Their Brain Awake

Many birds keep one eye open while sleeping, allowing them to rest and stay alert for predators at the same time.

Fact #68: Male Antechinuses Mate Until They Die

The male antechinus, a small marsupial, mates so intensely during the breeding season that it literally burns itself out and dies. Talk about dedication.

Fact #69: Crickets Have Ears on Their Knees

Crickets hear through tiny holes on their front legs, proving that nature loves to mix things up.

Fact #70: Tardigrades Can Survive in Space

Tardigrades, also known as water bears, can survive extreme conditions, including the vacuum of space. They're basically indestructible.

Fact #71: Parrots Can Laugh Together

Parrots are social animals, and when one laughs, the others often join in. It's like a feathery comedy club.

Fact #72: Spiders Can Balloon Across Oceans

Some spiders release silk threads that let them travel thousands of miles, using wind currents to "balloon" over the sea.

Fact #73: Ducks Can Change Gender

If a female duck loses her ovaries, she can develop male characteristics and behaviors. Nature doesn't play by the rules.

Fact #74: Hippos Don't Actually Swim

Despite spending most of their time in water, hippos can't swim. They walk or run along the riverbed instead.

Fact #75: Polar Bears Overheat Easily

Polar bears may thrive in the Arctic, but their thick fur and fat layers can cause them to overheat if they move too much.

Fact #76: Frogs Can Breathe Through Their Skin

In addition to their lungs, frogs absorb oxygen through their skin, especially when underwater.

Fact #77: Horses Can Sleep Standing Up

Horses have a special locking mechanism in their legs that lets them sleep standing up without falling over.

Fact #78: Chickens Can Dream

Chickens experience REM sleep, meaning they probably dream. Imagine a chicken dreaming about a bigger coop or extra snacks.

Fact #79: Reindeer Eyes Turn Blue in Winter

Reindeer eyes change color with the seasons, turning blue in winter to help them see in low light. Nature's built-in night vision.

Fact #80: Monkeys Use Millipedes as Bug Spray

Some monkeys rub millipedes on their fur because the insects release chemicals that repel mosquitoes.

Fact #81: Penguins Can Shoot Their Poop Like a Cannon

Penguins have a unique way of staying clean—they shoot their poop away from their nests with surprising force. Scientists even measured the speed for fun!

Fact #82: Giraffes Clean Their Ears With Their Tongues

Giraffes' tongues are so long (up to 20 inches) that they can use them to clean their ears. That's some serious multitasking.

Fact #83: Frogs Use Their Sticky Tongues Like a Slingshot

A frog's tongue is so sticky and elastic that it can catch insects in less than the blink of an eye. Dinner has never been so fast!

Fact #84: Cats Can't Taste Sweet Things

Unlike humans, cats lack the taste receptors for sweetness. So, no, they're not stealing your cake because it tastes good—they're just curious jerks.

Fact #85: Elephants Are Afraid of Bees

Despite their massive size, elephants will avoid bees at all costs. Some farmers even use beehives to keep them away from crops.

Fact #86: Parrots Can Live Over 80 Years

Some parrots, like the macaw, can outlive humans. Owning one isn't just a commitment—it's a lifetime partnership.

Fact #87: Baby Koalas Eat Their Mom's Poop

To help develop their digestion for eucalyptus leaves, baby koalas snack on their mom's poop. It's gross but apparently nutritious.

Fact #88: Sea Turtles Cry to Stay Hydrated

Sea turtles "cry" salty tears to rid their bodies of excess salt. It looks dramatic, but they're just staying healthy.

Fact #89: Birds Can Sleep While Perched

Birds' feet automatically lock onto branches while they sleep, keeping them from falling off. Talk about a built-in safety mechanism.

Fact #90: A Group of Owls is Called a Parliament

Owls are so wise-looking that their collective name is a "parliament." Who wouldn't trust them to make laws?

Fact #91: Bees Get Buzzed on Caffeine

Bees are more likely to visit plants with caffeine in their nectar. Apparently, even insects need their coffee fix.

Fact #92: Dolphins Give Each Other Nicknames

Dolphins not only use signature whistles to call each other but also develop nicknames for their pod members.

Fact #93: Some Fish Fart to Communicate

Herring use bubbles from their rear ends to send messages to each other. Scientists even call it Fast Repetitive Ticks, or FRTs. Yes, really.

Fact #94: Platypuses Glow Under UV Light

Under ultraviolet light, platypuses' fur glows green and blue. Nature really went all-out designing these creatures.

Fact #95: Squirrels Adopt Orphaned Babies

If a baby squirrel loses its mother, nearby squirrels will often adopt it and raise it as their own. Squirrels, the underrated heroes of the animal world.

Fact #96: Moths Get Drunk on Fermented Fruit

Moths are attracted to fermented fruit, which can make them drunk. Ever seen a moth flying erratically? It might be tipsy.

Fact #97: Turtles Can Talk Before They Hatch

Baby turtles communicate with each other through sounds while still in their eggs, coordinating when to hatch together for safety.

Fact #98: Hippos Spin Their Tails While Pooping

Hippos use their tails to fling poop around, marking their territory. It's messy, hilarious, and surprisingly effective.

Fact #99: Crocodiles Swallow Stones

Crocodiles eat rocks to help them stay balanced and digest food. Call it nature's version of a weight inside a keelboat.

Fact #100: Sloths Can Hold Their Breath Longer Than Dolphins

Sloths may be slow, but they're breath-holding champs, staying underwater for up to 40 minutes. Dolphins, by comparison, max out at about 10 minutes.

POP QUIZ

Ready to test your knowledge of the animal kingdom's funniest and quirkiest facts? Let's see how many of these you remember!

Questions

1. **True or False:** Wombats have cube-shaped poop to mark their territory.
2. Which animal accidentally grabs its own arm, mistaking it for a tree branch?
3. **True or False:** Penguins propose to their mates using pebbles.
4. What unique feature helps camels protect their eyes during sandstorms?
5. **True or False:** Sea otters hold hands while sleeping to avoid drifting apart.
6. Which animal spins its tail while pooping to mark its territory?
7. **True or False:** Flamingos' knees bend backward.
8. What amphibian throws up its stomach to clean it out when needed?
9. Which animal cries to remove excess salt from its body?
10. **True or False:** A platypus glows under UV light.

Answers

1. **True**: Wombats' cube-shaped poop prevents it from rolling away, making it perfect for marking their territory.
2. **Sloths**: Sloths sometimes grab their own arms instead of branches and fall from trees. Oops!
3. **True**: Male penguins give their mates a pebble as a proposal gift. Romantic, right?
4. **Three sets of eyelashes**: Camels have three layers of eyelashes to shield their eyes from blowing sand.
5. **True**: Sea otters hold hands while they sleep to stay together in the water. Cute and practical!
6. **Hippos**: Hippos spin their tails to fling poop around, marking their territory in a messy but effective way.
7. **True**: What looks like a flamingo's backward knee is actually its ankle. Their real knees are hidden!
8. **Frogs**: Frogs vomit their stomachs to clean them out. It's extreme, but it works.
9. **Sea turtles**: Sea turtles shed tears to remove excess salt from their bodies.
10. **True**: Under UV light, a platypus's fur glows blue and green. Nature loves its surprises!

How did you do? Whether you aced it or learned something new, the animal kingdom is full of surprises!

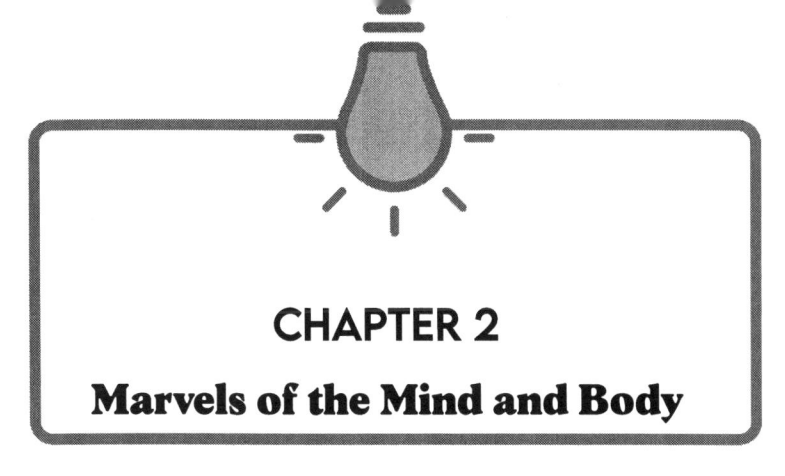

CHAPTER 2
Marvels of the Mind and Body

The human body is full of surprises, quirks, and hidden abilities. From rare conditions where people wake up speaking foreign accents to the brain's ability to trick you into tasting colors, humans are endlessly fascinating. This chapter explores the strange and unexpected powers of the mind and human body, revealing why we might just be the most peculiar species of all.

Fact #101: Some People Have an Extra Limb in Their Mind

A condition called Supernumerary Phantom Limb makes people feel like they have an extra arm or leg, even though it doesn't exist. Some even feel they can "move" it. Talk about a ghost limb upgrade!

Fact #102: Fear of Peanut Butter Sticking to the Roof of Your Mouth Exists

Arachibutyrophobia is the fear of peanut butter sticking to the roof of your mouth. While it might sound ridiculous, for some, it's the ultimate nutty nightmare!

Fact #103: Your Brain Can Trick You Into Hearing Your Name in a Crowd

It's called the "cocktail party effect." Even if 50 people are talking at once, your brain tunes in when it hears your name. It's basically a superhero power for nosy eavesdroppers.

Fact #104: You're Taller in the Morning

After a good night's sleep, your spine decompresses, making you about half an inch taller in the morning. By bedtime, gravity has squished you back down again. Thanks, Earth.

Fact #105: Your Feet Sweat Buckets

Each foot has about 250,000 sweat glands that can produce half a pint of sweat per day. That's a lot of sock laundry.

Fact #106: Your Brain Never Stops Talking to Itself

Your inner monologue is like your brain's own podcast. Fun fact: if you're arguing with yourself, you're also your own worst guest.

Fact #107: You Blink More Times Than a Traffic Light Changes

The average person blinks about 20 times per minute, or over 28,000 times a day. It's like your eyelids are doing mini workouts without you noticing.

Fact #108: Your Tongue Knows What Everything Feels Like

Your brain can imagine the texture of pretty much anything— like sandpaper or velvet—just by looking at it. And now you're trying not to imagine licking your keyboard. Too late!

Fact #109: Some People Can't Recognize Their Own Reflection

A condition called Mirror Agnosia causes people to lose the ability to understand mirrors. They might reach into the mirror to grab objects, thinking they're real. It's like getting trapped in a funhouse forever!

Fact #110: You Can't Breathe and Swallow at the Same Time

Don't try this now! Humans can't breathe and swallow simultaneously because there's only one lane for both. Nature didn't think this one through.

Fact #111: Your Fingernails Grow Faster When You're Stressed

Fingernails grow faster when you're stressed. It's like they're trying to claw their way out of the situation too!

Fact #112: Your Brain Runs on the Same Power as a Lightbulb

Your brain uses about 20 watts of energy, which is roughly what it takes to light a dim bulb. So the next time someone calls you "bright," you know why.

Fact #113: Humans Have Glow-in-the-Dark Skin

Your skin emits a faint glow, but it's 1,000 times weaker than what the human eye can see. Basically, we're all bioluminescent superheroes—but lame ones.

Fact #114: Your Stomach Has a Lining That Regenerates Every Few Days

If it didn't, your stomach acids would digest itself. So your body constantly refreshes the lining, saving you from becoming your own snack.

Fact #115: You Produce Enough Saliva to Fill Two Bathtubs a Year

Gross but true. That's a whole lot of drool—and it's just another reason to stay hydrated!

Fact #116: There's a Condition Where People Are Constantly "Drunk"

People with auto-brewery syndrome have bodies that ferment carbohydrates into alcohol, making them feel drunk even without drinking. Bread and pasta might give them the equivalent of a margarita buzz. It's fascinating—but definitely not fun to live with!

Fact #117: Your Nose Remembers Over 50,000 Smells

Your nose is like a super-organizer, cataloging tens of thousands of smells. Too bad it can't help you find your car keys.

Fact #118: Hiccups Are Mini Brain Malfunctions

Hiccups are caused by your diaphragm spasming, but why? No one really knows. Thanks, body, for this totally pointless feature.

Fact #119: Your Teeth Are Harder Than Steel

Tooth enamel is the hardest substance in your body. You could bite through steel—if you're okay with losing all your teeth afterward.

Fact #120: You Have Enough Blood Vessels to Wrap Around Earth

If you stretched out all your blood vessels, they'd cover about 60,000 miles. That's enough to circle the globe—twice. Too bad they don't come with frequent flyer miles.

Fact #121: Your Brain Has a "Delete" Button

Your brain deletes old, unused memories to make room for new ones. So if you forgot where you put your keys, maybe it's because your brain decided that TikTok dance was more important.

Fact #122: Some People Remember Every Single Day of Their Lives

A condition called hyperthymesia allows people to recall nearly every day of their life in detail. Imagine remembering exactly what you had for breakfast on a random Tuesday ten years ago—cool and exhausting!

Fact #123: Your Heart Skips a Beat When You Sneeze

Sneezing briefly alters your heart's rhythm, but don't worry—it's completely normal. Your body just likes to add a little drama.

Fact #124: The Brain Has No Pain Receptors

Your brain can't actually feel pain, which is why surgeons can perform brain surgery while you're awake. Good thing headaches come from somewhere else, or that'd be awkward.

Fact #125: Your Hair Knows the Weather

Your hair swells up in humid weather, which is why frizz happens. Turns out, your hair has a better weather app than your phone.

Fact #126: Your Brain is a Lazy Genius

The brain uses shortcuts to process information faster, which is why you can read mixed-up sentences like "Yuo cna raed tihs!" It's a smarty-pants, but it's also cutting corners.

Fact #127: Some People See Sounds

A condition called synesthesia causes people to hear colors or see sounds. Imagine hearing a song and literally seeing rainbows—an actual superpower.

Fact #128: Your Belly Button is a Jungle

Your belly button is home to thousands of bacteria, some unique to you. Basically, your navel is its own tiny ecosystem.

Fact #129: Laughter is Contagious

Ever heard someone laugh so hard you started laughing too? That's because your brain has "mirror neurons" that copy emotions. Laughter is basically nature's ultimate party trick.

Fact #130: Some People Can't Picture Things in Their Minds

People with a condition called aphantasia can't visualize images in their heads. If someone says, "Picture a beach," they think, "What picture?"

Fact #131: Your Brain Runs on Dreams

Dreams are like your brain's nightly exercise. It's thought that dreams help your brain process emotions and problem-solve. So if you dream of flying tacos, maybe your brain's working on dinner plans.

Fact #132: Goosebumps Have a Cool History

Goosebumps evolved to help our ancestors look bigger and scarier to predators. Now they just pop up during horror movies or power ballads.

Fact #133: Your Bones Are Always Changing

Your skeleton completely regenerates itself about every 10 years. So, technically, your bones are younger than you think.

Fact #134: The Brain Can Trick Itself Into Feeling Fake Pain

Ever heard of phantom limb syndrome? People who lose limbs can still feel sensations, like itching or pain, where their limb used to be. It's proof of how powerful the brain is.

Fact #135: Your Nose Never Forgets

Smell is the strongest sense tied to memory. A single whiff of grandma's cookies can take you straight back to childhood.

Fact #136: Hiccups Can Last for Years

The longest case of hiccups lasted 68 years. Talk about patience—and a really awkward dinner guest.

Fact #137: Some People Fear Vegetables

Lachanophobia is the fear of vegetables. Yes, it's real, and no, you can't use it as an excuse to skip eating your greens.

Fact #138: The Tongue is Stronger Than It Looks

Your tongue is one of the strongest muscles in your body for its size. You've probably already flexed it just reading this.

Fact #139: Your Eyes See the World Upside Down

Your eyes capture images upside down, and your brain flips them right side up. It's like having a built-in photo editor.

Fact #140: You're Always Breathing Through One Nostril

Your nostrils take turns working throughout the day. The other one is on break—because even noses need downtime.

Fact #141: People Can Forget Entire Languages

A rare condition called language attrition causes people to lose fluency in a language they once spoke perfectly. Imagine suddenly forgetting your favorite slang!

Fact #142: Tears Have Layers

Tears aren't just salty water—they're made of oils, mucus, and enzymes. Basically, your body created the world's most dramatic recipe.

Fact #143: Humans Are Biased Towards Faces

Your brain is so wired to recognize faces that it can see them in random objects, like clouds or your toast. Hello, Jesus bagel!

Fact #144: Your Body Is Always Fighting Tiny Battles

Every second, your immune system is fighting off bacteria and viruses. You're basically a walking Marvel movie.

Fact #145: Some People Can't Recognize Faces

Prosopagnosia, or face blindness, makes it hard for some people to recognize even their closest friends. They might recognize people by their hair or voice instead.

Fact #146: The Brain Can Multitask, But It's Terrible at It

Your brain switches between tasks instead of doing them simultaneously. So if you feel bad at multitasking, it's not you—it's science.

Fact #147: Some People Think They're Always Being Watched

A condition called Truman Show Syndrome causes people to believe their lives are part of a TV show. It's rare but fascinating.

Fact #148: Tongue Prints Are Unique

Just like fingerprints, your tongue has a one-of-a-kind pattern. Forget fingerprint scanners—imagine tongue scanners in the future!

Fact #149: Hair and Nails Don't Grow After Death

It's a myth that hair and nails grow after death—it's just the skin shrinking. Science ruins the spooky again!

Fact #150: You Can't Tickle Yourself

Your brain ignores your own touch to avoid confusion, so no matter how hard you try, you can't tickle yourself. Nice try, though.

Fact #151: Some People Wake Up With Foreign Accents

A rare condition called Foreign Accent Syndrome can cause people to wake up speaking with a completely different accent—often after surgery or an injury. Imagine going to bed sounding American and waking up talking like you're straight out of a Harry Potter movie!

Fact #152: Your Body Has Gold in It

The human body contains trace amounts of gold, mostly in your blood. Don't get any ideas—mining yourself won't make you rich.

Fact #153: The Human Brain Shrinks as You Age

Starting in your 30s, your brain begins to shrink slightly every year. But don't worry—it's still more powerful than the latest smartphone.

Fact #154: Some People Believe They're Dead

Cotard's Syndrome makes people believe they're dead or missing parts of their bodies. It's rare, fascinating, and proof that the brain is full of surprises.

Fact #155: Yawning is Contagious, Even for Dogs

Humans aren't the only ones who "catch" yawns—dogs do, too. If you yawn in front of your pup, there's a good chance they'll join in.

Fact #156: Human Teeth Are as Strong as Shark Teeth

Our teeth may look less scary, but they're just as strong as a shark's. The only difference? Sharks get unlimited replacements, and we don't.

Fact #157: Some People Laugh in Their Sleep

Ever heard of sleep laughter? It's like sleep talking but with giggles. No one knows why it happens, but it's probably harmless—and hilarious.

Fact #158: Your Bones Are Five Times Stronger Than Steel

Pound for pound, human bones are stronger than steel. Too bad they're not as shiny or good for building bridges.

Fact #159: Fingernails Grow Faster on Your Dominant Hand

Your dominant hand's fingernails grow slightly faster because it gets more blood flow. Yet another reason to favor one hand over the other.

Fact #160: The Brain Processes Images Faster Than You Blink

Your brain can identify images in as little as 13 milliseconds. It's like having the fastest photo app built in.

Fact #161: You Share 60% of Your DNA With Bananas

Humans and bananas share more than half their genetic makeup. Guess that explains why we're so a-peeling.

Fact #162: Some People Get Lost in Their Own Neighborhoods

Topographical disorientation makes it hard for people to recognize places they've been before, even their own street. GPS to the rescue!

Fact #163: The Brain Can Rewire Itself

The brain is super flexible and can rewire itself after injuries. It's like upgrading a computer while it's still running.

Fact #164: Your Ears and Nose Never Stop Growing

While the rest of your body stops growing, your ears and nose keep going. Now you know why Grandpa's ears look bigger than ever.

Fact #165: You Shed Millions of Skin Cells Per Day

Your body sheds about 30,000 to 40,000 skin cells every minute. That's millions of cells a day. Congratulations—you're a walking snow globe.

Fact #166: Your Body Has More Bacteria Than Cells

For every human cell in your body, there are about 10 bacteria hanging out with you. Think of it as a tiny, invisible fan club.

Fact #167: Some People Are Born Without Fingerprints

A rare genetic condition called adermatoglyphia leaves people without fingerprints. They might make terrible detectives, but great mystery villains.

Fact #168: Your Brain Can Play Tricks on You After a Limb is Amputated

Phantom limb syndrome can make amputees feel like their missing limb is still there. It's proof of just how strong the brain's connection to the body is.

Fact #169: Human Babies Grow Mustaches in the Womb

Babies grow a fine layer of hair, including a mustache, while in the womb. Don't worry—they shed it before birth and swallow it. Yikes.

Fact #170: Your Bones Are Full of Holes

Bones aren't solid—they're like sponges with tiny holes inside. That's why they're strong but lightweight enough to carry around.

Fact #171: Some People Can Remember Their Dreams in Extreme Detail

People with REM sleep behavior disorder act out their dreams and often remember them vividly. Imagine waking up and saying, "So, about that time I fought off a dragon…"

Fact #172: Your Eyes Can See UV Light—But Your Brain Filters It Out

If the lens of your eye is removed (don't try this), you'd be able to see ultraviolet light. Your brain has its reasons for filtering it, though—probably to avoid constant headaches.

Fact #173: You Can Hear Your Own Blood Flow

If you've ever heard a whooshing sound in a quiet room, that's your blood flowing through your ears. Turns out, your body loves ASMR.

Fact #174: Some People Are Literally Colorblind to Faces

A rare condition called facial color agnosia means some people can see faces, but not the natural colors in them. Imagine watching a black-and-white movie of everyone around you.

Fact #175: Your Hands and Feet Are Extremely Sensitive

About a quarter of your brain's sensory receptors are dedicated to your hands and feet. That's why a tiny tickle on your foot feels like an alarm.

Fact #176: Your Tongue is Covered in 8,000 Taste Buds

Your tongue's taste buds are grouped into clusters, each with about 100 cells. That's 8,000 little flavor interpreters working overtime.

Fact #177: Some People Can't Smell Anything

Anosmia is the inability to smell, and it can be genetic or caused by illness. It's like watching a movie on mute—still interesting, but not the full experience.

Fact #178: Your Brain is Always Predicting the Future

Your brain constantly predicts what will happen next, like a mental fortune-teller. That's how it helps you catch a ball—or anticipate when your cat is about to pounce.

Fact #179: Your Body Glows Brighter in the Afternoon

Humans emit more bioluminescence in the afternoon than in the morning or evening. Too bad we can't see it—we'd look like tiny suns.

Fact #180: Your Fingertips Are More Sensitive Than You Realize

Your fingertips can feel objects as small as 13 nanometers. That's 7,500 times thinner than a human hair. Who needs a magnifying glass?

Fact #181: Your Brain Prioritizes Faces

The brain has a special area, the fusiform face area, dedicated to recognizing faces. That's why you see smiley faces in pancakes, clouds, and random objects.

Fact #182: You're Hardwired to Love Sugar

Your brain lights up like fireworks when you eat sugar. It's an evolutionary trait that once helped us survive. Now it just helps us find dessert irresistible.

Fact #183: You're Constantly Replacing Your Stomach Lining

Your stomach lining replaces itself every 4-5 days to avoid being digested by stomach acid. Talk about a self-maintenance program.

Fact #184: You Shed Tears of Joy and Sadness

Your emotional tears contain stress hormones, which may explain why a good cry feels so cathartic. Tears are just nature's way of saying, "Let it out."

Fact #185: Some People Can't Feel Pain

Congenital insensitivity to pain makes people unable to feel physical pain. While it sounds like a superpower, it can actually be dangerous—pain is your body's alarm system.

Fact #186: Your Brain is the Fattest Organ

The human brain is about 60% fat, making it the fattiest organ in your body. You've literally got a fat head—and it's a good thing.

Fact #187: You Can't Sneeze With Your Eyes Open

Sneezing causes your body to automatically close your eyes. It's nature's way of making sure you don't accidentally eject your eyeballs. Thanks, evolution!

Fact #188: Your Hair Can Reveal Your Stress Levels

Hair retains a record of your stress hormones, like a time capsule of all your meltdowns. No wonder it turns gray!

Fact #189: Your Body Has a Built-in Detox System

Forget juice cleanses—your liver and kidneys are working 24/7 to detox your body. Turns out, your organs are the real MVPs.

Fact #190: Your Body Has Hidden Reflexes

Ever notice your arm or leg twitch suddenly when you're falling asleep? It's called a hypnic jerk, and it's your body's way of checking you're still alive. Thanks for the scare, brain!

Fact #191: Some People Hear Voices—And It's Not Always Bad

Auditory hallucinations aren't always negative. Some people hear friendly voices that provide comfort or even inspiration. The brain works in mysterious ways.

Fact #192: Your Ears Never Stop Hearing

Your ears are always "on," even when you sleep. Your brain just decides not to bother you with the noise.

Fact #193: Your Body is Mostly Water

About 60% of your body is water. That means you're basically a walking water balloon.

Fact #194: Your Heart Beats Over 100,000 Times a Day

That's more than 35 million beats a year. Imagine working that hard and never getting a coffee break.

Fact #195: Your Nose is a Heating System

Your nose warms up cold air before it reaches your lungs. It's like having a built-in furnace in your face.

Fact #196: Your Blood Travels 12,000 Miles a Day

If your blood vessels were roads, your heart would be driving around the world every single day.

Fact #197: You Can Trick Your Brain Into Thinking a Fake Hand is Yours

The "rubber hand illusion" shows how easily your brain can be fooled into feeling ownership over a fake hand. The mind is wild.

Fact #198: You Blink Less When You're Focused

When concentrating on something (like reading this book), you blink less often. Your eyeballs are doing overtime.

Fact #199: Some People Think They're Made of Glass

A rare condition called The Glass Delusion makes people believe they are fragile and could shatter at any moment. It was most common in the Middle Ages, and even King Charles VI of France believed he was made of glass!

Fact #200: You Can Smile in 19 Different Ways

Your face has 43 muscles, and you can use them to create at least 19 different types of smiles. Whether it's a cheesy grin or a mischievous smirk, you're a smiling pro.

POP QUIZ

Let's see how much you remember about the weird and wonderful quirks of the human mind and body. Get ready to flex your brainpower with these 10 fun questions!

Questions

1. **True or False:** The human body sheds about 9 pounds of skin every year.
2. What rare condition makes people wake up speaking with a different accent?
3. **True or False:** Your stomach replaces its lining every few days to avoid digesting itself.
4. What surprising material makes up 60% of the human brain?
5. **True or False:** Humans share over 50% of their DNA with bananas.
6. What happens to your brain as you age—it grows, shrinks, or stays the same size?
7. **True or False:** Phantom limb syndrome can make amputees feel sensations in a missing limb.
8. What is the name of the condition where people can't recognize their own reflection?
9. What is the name of the condition where people can hear colors and see sounds?
10. How many times does your heartbeat in a single day?

Answers

1. **True**: Humans shed about 9 pounds of skin every year, which contributes to a lot of household dust.
2. **Foreign Accent Syndrome**: This rare condition can make people suddenly speak with a new accent, often after surgery or injury.
3. **True**: Your stomach constantly regenerates its lining to protect itself from being dissolved by its own acid.
4. **Fat**: The brain is 60% fat, making it the fattiest organ in the body. It's literally full of good fuel!
5. **True**: Humans and bananas share over half of their DNA, proving that we have more in common with fruit than we thought.
6. **Shrinks**: Your brain begins shrinking in your 30s. It's subtle, but the process continues as you age.
7. **True**: Phantom limb syndrome causes people to feel sensations, like pain or itching, in a limb that's no longer there.
8. **Mirror Agnosia**: This condition makes people lose the ability to understand mirrors, leading them to treat reflections like real objects.
9. **Synesthesia**: A condition where people experience sensory overlap, such as hearing colors or seeing sounds, making sensory experiences blend together.
10. **100,000 times**: Your heart beats over 100,000 times a day, pumping blood through your body like the hardworking machine it is.

How did you score? Whether you nailed it or learned something new, the human body is full of surprises!

CHAPTER 3

Funny and Weird Facts
About Pop Culture

Pop culture isn't just about catchy songs, blockbuster movies, and viral memes—it's a treasure trove of bizarre, hilarious, and downright spooky stories. From behind-the-scenes movie mishaps to weird celebrity quirks, this chapter dives into the strangest corners of the entertainment world.

Fact #201: A Movie Once Sent People Running Out of Theaters

When *The Exorcist* premiered in 1973, some audience members fainted, vomited, or ran out of the theater screaming. One viewer even sued the movie's creators, claiming it caused a miscarriage. Now that's some serious horror!

Fact #202: The Wizard of Oz Nearly Cooked Its Tin Man

In *The Wizard of Oz*, the original Tin Man's makeup was made with aluminum powder, which poisoned actor Buddy Ebsen. The studio replaced him, but even the new Tin Man had to wear makeup so hot it left burns. The real villain of Oz? Bad special effects.

Fact #203: A Song Made by Accident Became a Huge Hit

The theme for *The X-Files* was created by mistake when composer Mark Snow's elbow hit his keyboard. The eerie sound became iconic—and proves even accidents can be spooky.

Fact #204: Horror Movies Can Burn Calories

Watching a scary movie can make your heart race, burning up to 200 calories. So, technically, binge-watching horror flicks could count as cardio. Who needs a treadmill when you have *Halloween*?

Fact #205: Barbie Has Had More Careers Than Most People

Since her debut in 1959, Barbie has had over 200 careers, including astronaut, surgeon, and even presidential candidate. Honestly, Barbie might have a better résumé than anyone you know.

Fact #206: The Simpsons Predicted Smartwatches

In a 1995 episode of *The Simpsons*, Lisa's future fiancé wore a watch that let him make phone calls. Nearly two decades later, smartwatches became a reality. Coincidence, or is Matt Groening a time traveler?

Fact #207: Shrek Was Voiced in a Swamp-Like Recording Studio

Mike Myers recorded many of Shrek's lines in a studio surrounded by mud and swamp props to "get into character." Dedication level: ogre.

Fact #208: A Real Skeleton Was Used in Poltergeist

To save money, the filmmakers used real human skeletons for the pool scene in *Poltergeist*. Nobody told actress JoBeth Williams until afterward. Talk about an unexpected scare.

Fact #209: Some People Have a Fear of Mickey Mouse

Mousephobia is a real thing, and for some people, the sight of Mickey Mouse in a theme park is the stuff of nightmares. Disney World isn't magical for everyone!

Fact #210: A Voice Actor Played 30 Characters in One Show

Mel Blanc, the legendary voice actor, voiced nearly all the characters in *Looney Tunes*, including Bugs Bunny, Daffy Duck, and Porky Pig. Basically, he was a one-man cartoon studio.

Fact #211: Ghosts Stalk the Harry Potter Set

The *Harry Potter* movies were filmed at England's Leavesden Studios, which is rumored to be haunted. Cast members reported seeing strange figures on set. Maybe they were just extra extras?

Fact #212: A Doll Sparked Real-Life Fear

Annabelle, the creepy doll from *The Conjuring* movies, is based on a real haunted Raggedy Ann doll. It's locked up in a glass case in a museum—where it's hopefully staying forever.

Fact #213: Friends Almost Had a Totally Different Cast

Before the final casting, other actors were considered for the roles of Ross, Rachel, and Joey. Imagine *Friends* without Jennifer Aniston or Matt LeBlanc—it'd be a very different sitcom.

Fact #214: A Movie Prop Became a National Treasure

The Maltese Falcon from the 1941 movie of the same name sold at auction for $4 million. It's literally worth its weight in gold—well, almost.

Fact #215: A TV Broadcast Was Hijacked by a Creepy Masked Figure

In 1987, two Chicago TV stations were interrupted by someone wearing a Max Headroom mask. The eerie broadcast showed the figure speaking nonsense and laughing maniacally before it abruptly ended. The culprit was never caught.

Fact #216: Jaws Was a Total Disaster—At First

The mechanical shark in *Jaws* broke down constantly, forcing Steven Spielberg to shoot scenes where the shark wasn't visible. The result? A scarier movie that became a classic. Sometimes, accidents pay off.

Fact #217: Celebrities Sometimes Have Weird Last Wishes

Tupac Shakur's friends mixed his ashes with marijuana and smoked them, fulfilling one of his final requests. If that's not bizarre enough, Gene Roddenberry, the creator of *Star Trek*, had his ashes launched into space.

Fact #218: E.T. Was Voiced by a Woman Who Smoked 2 Packs a Day

E.T.'s raspy voice was created by Pat Welsh, whose chain-smoking gave her the perfect alien tone. Proof that bad habits can sometimes land you an iconic role.

Fact #219: A Disney Movie Was Cursed With Production Troubles

Disney's *The Emperor's New Groove* had so many issues during production that it spawned a documentary called *The Sweatbox*, showing just how chaotic it was. Thankfully, the final movie was hilarious.

Fact #220: The First Star Wars Movie Nearly Flopped

Before *Star Wars: A New Hope* hit theaters, the cast thought it would fail. Even George Lucas wasn't confident. Turns out, a galaxy far, far away was closer to greatness than anyone realized.

Fact #221: Michael Myers' Mask in Halloween Was a Repurposed Captain Kirk Mask

The terrifying mask worn by Michael Myers in *Halloween* was originally a cheap Captain Kirk mask from *Star Trek*. The filmmakers painted it white, cut the eye holes bigger, and created a horror icon.

Fact #222: Psycho Was the First Movie to Show a Toilet Flushing

Alfred Hitchcock's *Psycho* broke taboos by including a flushing toilet on screen. Audiences were shocked—and not just by the shower scene.

Fact #223: A Parrot Helped Write Pirates of the Caribbean

Writer Ted Elliott got the idea for Jack Sparrow's character from his pet parrot, who had a habit of swaggering around like it owned the place. No word on whether the parrot got any royalties.

Fact #224: Toy Story 2 Was Almost Deleted

A computer glitch nearly erased *Toy Story 2* during production. Luckily, a backup copy saved the movie from disappearing into the void, much to Buzz and Woody's relief.

Fact #225: A Ghost Is Rumored to Haunt The Conjuring Movie Set

During filming for *The Conjuring*, cast and crew claimed to feel sudden cold spots, hear strange noises, and even see shadowy figures. Spooky coincidences, or something more?

Fact #226: There's a Secret Disney Movie Vault

Disney keeps a literal vault of their classic movies, locking them away for years before re-releasing them. It's like a treasure chest of nostalgia, guarded by corporate dragons.

Fact #227: James Cameron Drew Titanic's Sketches Himself

The famous nude sketch of Rose in *Titanic* was drawn by director James Cameron. Talk about a hands-on approach to filmmaking!

Fact #228: Movie Trailers Used to Play After the Movie

The term "trailer" comes from the fact that they originally played *after* movies. Of course, no one stayed to watch them, so they were moved to the beginning.

Fact #229: A Famous Movie Line Was Improvised

"Here's looking at you, kid," from *Casablanca*, was ad-libbed by Humphrey Bogart. Turns out, Hollywood's best moments aren't always in the script.

Fact #230: The Shining's Famous Door Scene Took 60 Tries

Jack Nicholson chopped through 60 doors for the iconic "Here's Johnny!" scene in *The Shining*. By the end, no doors in Hollywood were safe.

Fact #231: Keanu Reeves Gave Away His Matrix Earnings

Keanu Reeves reportedly gave away most of his *Matrix* earnings—around $75 million—to the movie's special effects and costume team. Neo might be a hero on screen, but Keanu is one in real life.

Fact #232: A Shark Exploded During Filming of Deep Blue Sea

During production, one of the animatronic sharks exploded on set. Luckily, it wasn't real—just an expensive, mechanical disaster.

Fact #233: A Celebrity Owns the World's Largest Smurf Collection

Neil Patrick Harris owns a massive collection of Smurf memorabilia. It's so big, he could probably open a theme park.

Fact #234: The Office Almost Didn't Happen

The American version of *The Office* was almost canceled after its first season due to low ratings. Thankfully, it stuck around to give us Dwight Schrute and countless awkward laughs.

Fact #235: The Blair Witch Project Was Sold as Real Footage

When *The Blair Witch Project* premiered, many viewers believed it was real. The marketing team even made fake missing-person posters to sell the story.

Fact #236: The Game of Thrones Set Caught Fire

During the filming of *Game of Thrones*, a special effects fire got out of control, destroying part of the set. At least the dragons weren't blamed.

Fact #237: A Harry Potter Scene Was Too Gross for Some Actors

In *The Chamber of Secrets*, the scene with Hagrid's giant spider, Aragog, made some cast members refuse to step on set. Spider-fearing fans definitely relate.

Fact #238: Jurassic Park's Dinosaur Sounds Were Borrowed from Tortoises

The terrifying roars in *Jurassic Park* weren't all that scary in real life—some were recordings of tortoises mating. Let that sink in next time you watch a T. rex.

Fact #239: Nicolas Cage Bought a Haunted Mansion

Nicolas Cage once bought the LaLaurie Mansion in New Orleans, one of the most haunted houses in America. Maybe he thought he could use it for acting inspiration.

Fact #240: Marlon Brando Wore Pajamas on the Set of The Godfather

Marlon Brando was so relaxed on the set of *The Godfather* that he sometimes showed up in his pajamas. When you're the Godfather, who's going to argue?

Fact #241: A Movie Was Shot Inside a Real Volcano

The movie *The Devil at 4 O'Clock* was partially filmed inside an active volcano. Thankfully, the cast and crew made it out without turning into barbecue.

Fact #242: Someone Wore a Monkey Suit in King Kong

The original *King Kong* (1933) wasn't all stop-motion magic—some scenes featured an actor in a monkey suit. Special effects have come a long way since then.

Fact #243: Arnold Schwarzenegger Got Paid to Say Only 17 Lines

In *The Terminator*, Arnold Schwarzenegger was paid $75,000 and had only 17 lines of dialogue, including "I'll be back." That's a big payday for minimal small talk.

Fact #244: Horror Movies Are Filmed in Freezing Temperatures

To make actors look genuinely scared, many horror movies, like *The Thing*, are filmed in freezing conditions. Nothing says "terrified" like chattering teeth.

Fact #245: A Music Video Once Shut Down New York City

Michael Jackson's *Bad* music video, directed by Martin Scorsese, caused such a frenzy it temporarily shut down parts of New York City. Who's bad? Apparently, Michael.

Fact #246: E.T. Loved Reese's Pieces by Accident

Reese's Pieces became famous after *E.T.* because M&M's turned down the chance to be featured in the movie. Big mistake, Mars Company.

Fact #247: Actors in Cats Didn't Understand the Plot

During interviews, several actors admitted they didn't understand the plot of *Cats*. Honestly, who did?

Fact #248: The Lion King Was Almost Called King of the Jungle

Disney considered naming *The Lion King King of the Jungle*, until someone realized lions don't live in jungles. Awkward.

Fact #249: A Bizarre Set Accident in The Dark Knight

During the filming of *The Dark Knight*, a truck accidentally crashed into the Batmobile. Bruce Wayne wasn't happy, but the insurance company sure was.

Fact #250: The Most Expensive Movie Prop Was a Spaceship

The miniature model of the Millennium Falcon used in *Star Wars* sold for $2.3 million at auction. Turns out, going to a galaxy far, far away isn't cheap.

Fact #251: The Blood in Psycho Was Chocolate Syrup

In Alfred Hitchcock's *Psycho*, the iconic shower scene used chocolate syrup as fake blood. It looked better in black-and-white—and probably smelled delicious.

Fact #252: A Stuntman Was Set on Fire for Real in Casino Royale

In *Casino Royale*, a stuntman was set on fire for over 45 seconds, setting a record for the longest controlled burn on film. Talk about a hot performance!

Fact #253: Horror Movie Screams Are Sometimes Borrowed from Other Movies

Many films recycle screams from older movies, like the Wilhelm scream, a classic sound effect heard in *Star Wars*, *Indiana Jones*, and more.

Fact #254: The Matrix Code Is a Sushi Recipe

The iconic green code in *The Matrix* is actually made up of Japanese characters from a sushi cookbook. Turns out, the Matrix is more delicious than dystopian.

Fact #255: A TV Show's Dog Got More Fan Mail Than the Cast

Lassie, the famous TV dog, received more fan mail than any of her human co-stars. Turns out, everyone loves a good doggo.

Fact #256: Horror Movie Fans Are Better at Handling Anxiety

Studies show that people who love horror movies are often calmer under stress. Maybe those jump scares are good training for real-life surprises.

Fact #257: Titanic Made People Seasick in Theaters

Some viewers got motion sickness from the ship-sinking scenes in *Titanic*. That's a new level of immersive filmmaking.

Fact #258: Tom Cruise Does Most of His Own Stunts

Tom Cruise performs many of his own stunts, including hanging off an airplane in *Mission Impossible: Rogue Nation*. It's hard to tell where Ethan Hunt ends and Tom Cruise begins.

Fact #259: The Office Cast Had Real Jobs in the Pilot Episode

To add realism, some cast members of *The Office* were assigned actual office tasks during filming, like making spreadsheets and answering phones.

Fact #260: The Highest-Grossing Movie Flopped in Some Countries

While *Avatar* is the highest-grossing movie of all time, it completely flopped in Japan, where audiences didn't connect with its story.

Fact #261: It's a Wonderful Life Used Soap for Snow

The snow in *It's a Wonderful Life* was created using soap and water. Real snow didn't look good on camera, so Hollywood made it squeaky clean.

Fact #262: Stranger Things Almost Had a Different Title

Before settling on *Stranger Things*, the creators considered calling the show *Montauk*. Imagine the Demogorgon in a beach setting!

Fact #263: The Blair Witch Project Cast Got Lost While Filming

The actors in *The Blair Witch Project* weren't just acting—they actually got lost in the woods during filming. Method acting at its finest.

Fact #264: The Godfather Used Real Mobsters

To add authenticity, the producers of *The Godfather* consulted real-life mobsters. Some even appeared as extras in the wedding scene.

Fact #265: Frozen Sparked a Baby Name Craze

After Disney's *Frozen* became a hit, the name "Elsa" saw a huge surge in popularity. Parents everywhere couldn't "let it go."

Fact #266: A Scene in The Avengers Was Shot in a Parking Lot

The post-credits scene where the Avengers eat shawarma was filmed in a random parking lot. Movie magic strikes again!

Fact #267: The Breaking Bad Meth Recipe is Incomplete on Purpose

To avoid inspiring real-life criminals, *Breaking Bad* deliberately left out key steps in its meth-making process. Science says: no copycats allowed.

Fact #268: The Wizard of Oz Was Filmed in Two Technicolor Shades

To make Dorothy's ruby slippers pop, *The Wizard of Oz* used a special version of Technicolor that made red appear extra vibrant. Fashion and film innovation!

Fact #269: Pulp Fiction's Suitcases Were Left Empty

Quentin Tarantino left the contents of the glowing suitcase in *Pulp Fiction* a mystery—even to the cast. Fans still debate what was inside.

Fact #270: Shrek Almost Starred Chris Farley

Before Mike Myers voiced Shrek, Chris Farley recorded nearly all the dialogue. After his untimely passing, the role was recast—but Farley's version is still legendary among fans.

Fact #271: Hitchcock Hypnotized an Actress on Set

While filming *Spellbound*, Alfred Hitchcock hypnotized actress Ingrid Bergman for a dream sequence. That's one way to get into character.

Fact #272: Star Wars' Lightsaber Sound Was Made Using a TV

The iconic lightsaber sound in *Star Wars* came from a mix of a buzzing TV and a microphone passing an electric motor. Sci-fi magic meets household appliances.

Fact #273: The Friends Couch Was Found in a Basement

The iconic orange couch from *Friends* was a lucky find—it was discovered in the Warner Bros. studio basement. Talk about thrifting success!

Fact #274: A Scene in Harry Potter Was Almost Cut Because of Laughter

The scene where Harry, Hermione, and Ron wear polyjuice potion disguises in *The Chamber of Secrets* was almost cut because the actors couldn't stop laughing at each other.

Fact #275: The Lord of the Rings Used 48,000 Fake Feet

The hobbits wore custom prosthetic feet, and over 48,000 pairs were used during filming. Imagine doing laundry for that many socks.

Fact #276: Horror Movies Sometimes Use Baby Monitors

Baby monitors are often used to create creepy sound effects in horror films. Next time you hear static, it might feel a little spookier.

Fact #277: Pixar Movies Always Hide Easter Eggs

Every Pixar movie contains hidden references to other Pixar films, creating a shared universe. Look closely, and you'll find Nemo in *Monsters, Inc.*

Fact #278: Darth Vader Was Voiced by a Different Actor

James Earl Jones provided Darth Vader's voice, but the man in the costume was played by bodybuilder David Prowse. Two actors, one terrifying villain.

Fact #279: A Snake Was Accidentally Set Loose on the Set of Raiders of the Lost Ark

During the filming of the snake scene, one slithered away from its enclosure and scared the crew. Hopefully, Indiana Jones wasn't around for that!

Fact #280: Bruce Lee's Famous Noises Were Added After Filming

Bruce Lee didn't make all those iconic fighting sounds during filming—they were added in post-production to amp up the action.

Fact #281: A Cat Was the Star of Alien

Jonesy the cat became a fan-favorite in *Alien* and even got its own stunt double. Cats truly run the galaxy.

Fact #282: A Single Line of Dialogue Saved Toy Story

During early test screenings, audiences thought Woody was too mean. Adding the line "You are a cool toy!" softened his character, saving the movie from becoming a flop.

Fact #283: Will Smith Turned Down The Matrix

Will Smith was offered the role of Neo in *The Matrix* but turned it down to star in *Wild Wild West*. He has since admitted it wasn't his best career move.

Fact #284: A Ghostbusters Scene Was Improvised

Bill Murray's famous line "He slimed me!" in *Ghostbusters* was ad-libbed. Murray's knack for improvisation made the scene an instant classic.

Fact #285: A Movie Killed the Disco Craze

The 1979 movie *Can't Stop the Music*, starring the Village People, is credited with ending the disco era. The movie flopped so hard it created a cultural backlash.

Fact #286: There's a Hidden Mickey in Every Disney Movie

Disney animators sneak Mickey Mouse silhouettes into all their movies. Some are easy to spot, while others are so well-hidden they're like a scavenger hunt.

Fact #287: A Famous Horror Movie Was Inspired by Real Events

The Texas Chainsaw Massacre was loosely inspired by real-life murderer Ed Gein. Thankfully, the chainsaw part was purely fictional.

Fact #288: Back to the Future Almost Used a Fridge Instead of a Car

Early drafts of *Back to the Future* had Marty traveling through time in a refrigerator. The DeLorean was a much cooler choice—pun intended.

Fact #289: The Lion King Had a Secret Shakespearean Twist

Disney's *The Lion King* is loosely based on Shakespeare's *Hamlet*. Both feature power struggles, family betrayal, and ghostly advice. Hakuna Matata meets "To be or not to be."

Fact #290: The Harry Potter Books Were Almost Published Under a Different Name

The publisher thought boys wouldn't read a book by a woman, so J.K. Rowling used her initials instead of her full name. Spoiler: boys read it anyway.

Fact #291: Daniel Radcliffe Was Allergic to His Harry Potter Glasses

The round glasses Daniel Radcliffe wore as Harry Potter caused an allergic reaction. Luckily, they swapped them out with a hypoallergenic pair.

Fact #292: Gremlins Inspired the PG-13 Rating

Gremlins was so scary for kids that it inspired the creation of the PG-13 rating. Apparently, cute furry creatures can give you nightmares.

Fact #293: A Chicken Almost Ruined The Godfather

During the wedding scene in *The Godfather*, a rogue chicken wandered into the shot and refused to leave. The filmmakers had to reshoot the scene multiple times.

Fact #294: A Disney Princess Movie Was Too Scary for Kids

When *Snow White and the Seven Dwarfs* premiered in 1937, many kids fled the theater in terror during the Queen's transformation scene. Who knew Disney could be so dark?

Fact #295: A Sandwich Inspired The Hitchhiker's Guide to the Galaxy

Douglas Adams came up with the idea for *The Hitchhiker's Guide to the Galaxy* while lying drunk in a field, clutching a sandwich. Talk about divine sandwich intervention.

Fact #296: A Celebrity Inspired the Voice of Dory in Finding Nemo

Ellen DeGeneres' voice on her talk show inspired Pixar to cast her as Dory. Her natural comedic timing was perfect for the forgetful fish.

Fact #297: Jurassic Park's Velociraptors Were Half the Size in Real Life

In real life, Velociraptors were about the size of turkeys. Spielberg made them bigger and scarier for the movie—and we're thankful he did.

Fact #298: The Shining Script Changed Daily

Stanley Kubrick constantly rewrote *The Shining* script during filming, often giving actors their lines minutes before shooting. Talk about keeping everyone on edge.

Fact #299: A Star Wars Prop Became a Real Scientific Tool

The lightsaber handle was made from an old Graflex camera flash handle. Years later, similar technology inspired real laser devices.

Fact #300: The Jaws Theme Almost Didn't Happen

The now-iconic two-note *Jaws* theme was dismissed as "too simple" by some producers. John Williams insisted, and it became one of the most recognizable scores in history.

POP QUIZ

Think you're a pop culture pro? Let's put your knowledge of movies, music, and all things entertainment to the test with these 10 fun questions!

Questions

1. **True or False:** The iconic green code in *The Matrix* is made up of sushi recipes.
2. What movie's horrifying pool scene used real human skeletons as props?
3. **True or False:** Reese's Pieces were featured in *E.T.* because M&M's turned down the chance.
4. Which actor originally recorded nearly all of Shrek's lines before the role went to Mike Myers?
5. **True or False:** The shower scene in *Psycho* used chocolate syrup for blood.
6. What Disney movie led to the creation of the PG-13 rating because it was too scary for kids?
7. **True or False:** Marty McFly was originally supposed to time travel using a refrigerator in *Back to the Future*.
8. What famous sci-fi movie almost failed because its mechanical shark kept breaking down?
9. **True or False:** The *Harry Potter* movies were filmed in a haunted studio.
10. What iconic horror villain wore a repurposed Captain Kirk mask?

Answers

1. **True:** The Matrix's green code came from a sushi cookbook's Japanese characters.
2. **Poltergeist:** Real skeletons were used in the pool scene without the actress knowing.
3. **True:** M&M's turned down E.T., making Reese's Pieces famous.
4. **Chris Farley:** He voiced Shrek before Mike Myers took over.
5. **True:** Psycho used chocolate syrup as blood in the shower scene.
6. **Gremlins:** Its intensity inspired the creation of the PG-13 rating.
7. **True:** Marty McFly's time machine was almost a refrigerator.
8. **Jaws:** A broken shark made Spielberg's scenes scarier.
9. **True:** Harry Potter was filmed at a studio rumored to be haunted.
10. **Michael Myers:** Halloween's villain wore a modified Captain Kirk mask.

How did you do? Whether you nailed it or learned something new, pop culture is full of surprises!

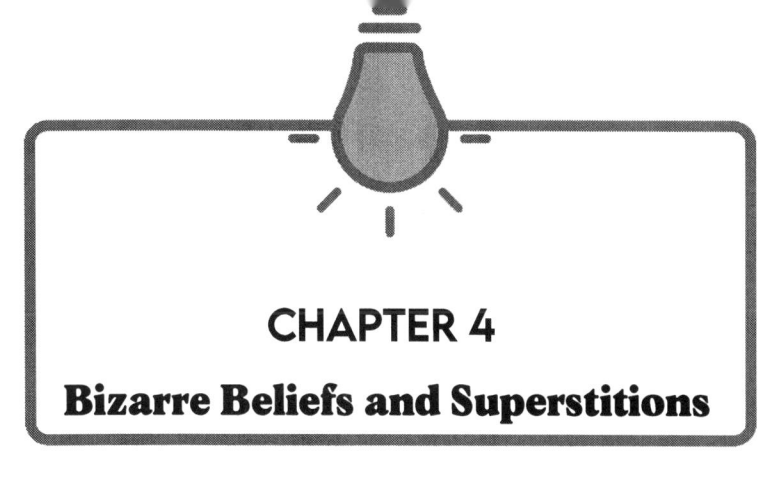

CHAPTER 4

Bizarre Beliefs and Superstitions

Beliefs and superstitions have been around forever, from lucky charms to downright creepy rituals. But let's be real—some of them are so weird, they'll leave you scratching your head, laughing, or maybe even a little spooked. Buckle up for a mix of funny, freaky, and mysterious superstitions from around the world!

Fact #301: Knocking on Wood Started With Tree Spirits

The habit of knocking on wood comes from ancient beliefs that spirits lived in trees. By knocking, people hoped to wake up the spirits for good luck. Today, we knock on wood just to avoid jinxing ourselves—or looking like fools.

Fact #302: Sneezing Was Once Considered Dangerous

In ancient times, people believed sneezing could expel your soul, leaving your body vulnerable to evil spirits. That's why we say "bless you." Turns out, it's not just polite—it's spiritual security.

Fact #303: Black Cats Are Lucky and Unlucky

In the Middle Ages, black cats were thought to be witches in disguise. But in Japan, they're seen as good luck. So, are they magical companions or furry omens? Depends on where you're standing!

Fact #304: The Number 13 Scares So Many People It Has Its Own Phobia

Triskaidekaphobia is the fear of the number 13, and it's so widespread that many buildings skip the 13th floor. Apparently, jumping straight to 14 makes it all less spooky.

Fact #305: Don't Whistle Indoors Unless You Want a Ghostly Guest

In Russia, whistling indoors is said to summon evil spirits—or at least bad luck. It's not worth the risk, so keep your *whistling skills* outdoors.

Fact #306: Breaking a Mirror Equals Seven Years of Bad Luck

This superstition comes from the ancient belief that mirrors contain a piece of your soul. Breaking one was thought to damage your spirit. Good thing mirrors aren't as fragile these days—hello, shatterproof glass!

Fact #307: Stepping on Cracks Really Freaked People Out

The rhyme "Step on a crack, break your mother's back" wasn't just for kids. It comes from an old belief that cracks symbolized gaps between the spiritual world and Earth. Creepy—and a great excuse to avoid sidewalk repairs.

Fact #308: Spilling Salt Summons the Devil

In medieval Europe, spilling salt was thought to invite the Devil into your life. The solution? Throwing a pinch of salt over your left shoulder to blind him. Who knew seasoning your food could be so risky?

Fact #309: Owls Are Seen as Harbingers of Death

In some cultures, hearing an owl's hoot at night is considered a bad omen, symbolizing death. Other cultures think they're wise messengers. Either way, hearing an owl at 3 a.m. will give anyone the creeps.

Fact #310: Tucking Your Thumbs Protects Your Parents

In Japan, tucking your thumbs into your fists when passing a graveyard is said to protect your parents from harm. Thumbs are considered "parent fingers," so better tuck them in just in case.

Fact #311: Friday the 13th Has an Unlucky Reputation

The superstition around Friday the 13th likely comes from Christianity, where 13 guests at the Last Supper and Jesus' crucifixion on a Friday combined into one seriously bad vibe. Today, it's just an excuse for horror movie marathons.

Fact #312: Holding Your Breath in Cemeteries

In some cultures, people hold their breath while passing a cemetery to avoid inhaling spirits. If you've ever wondered why someone suddenly stops talking on a graveyard drive, now you know.

Fact #313: The Evil Eye Is More Than a Trendy Jewelry Design

The evil eye, a curse said to be caused by envious looks, is believed to bring misfortune. People wear talismans to ward it off. Fashion and spiritual protection? Win-win.

Fact #314: Don't Trim Your Nails at Night

In South Korea, it's believed trimming your nails at night could attract rats who might steal your soul. Honestly, just the idea of rats is enough to save the nail care for daytime.

Fact #315: Don't Sleep With Your Feet Facing the Door

In feng shui, sleeping with your feet pointing toward the door is called the "coffin position" and is thought to bring bad luck. Maybe rearrange your furniture before bedtime, just in case.

Fact #316: Say Goodbye to Your Shoes on the Table

In many cultures, placing shoes on a table is a major no-no, thought to bring bad luck or even death. Plus, it's just plain gross.

Fact #317: Red Ink is Bad News

In South Korea, writing someone's name in red ink is believed to bring misfortune—or even death. Stick to blue or black pens if you're planning to stay friends.

Fact #318: Wishing on Eyelashes Started as a Protective Spell

Blowing an eyelash off your finger while making a wish dates back to medieval times. It was thought to ward off bad spirits. Turns out, your lashes are magical multitaskers.

Fact #319: An Open Umbrella Indoors Invites Bad Luck

This one likely started because open umbrellas are hazardous indoors, but some say it angers the spirits who protect your home. Better safe than poking someone in the eye—or worse.

Fact #320: Mirrors Can Trap Ghosts

In some cultures, it's believed that mirrors can trap spirits. Covering them after a death is meant to keep ghosts from getting stuck or roaming your home. Who knew mirrors could double as ghost traps?

Fact #321: Crossing Your Fingers for Luck Comes From Christianity

The habit of crossing your fingers started as a secret sign among early Christians to ward off evil and summon divine protection. Now, it's mostly used to wish for Wi-Fi to connect faster.

Fact #322: A New Broom Means New Beginnings

In some cultures, it's bad luck to bring an old broom to a new home because it's thought to carry bad energy. Time to upgrade your cleaning supplies when you move!

Fact #323: Whistling at Night Summons the Devil

In Turkey, it is believed that whistling at night can summon the devil. Better keep your tunes to yourself after sunset!

Fact #324: Yellow Flowers Can End Relationships

In Russia, giving yellow flowers to someone is considered a sign that a breakup or betrayal is coming. Better stick to roses—unless you're sending a message.

Fact #325: Walking Under a Ladder Has Dark Origins

This superstition dates back to ancient Egypt, where a ladder leaning against a wall created a triangle, considered sacred.

Walking through it was seen as breaking the harmony of the gods. Today, it's just a safety hazard.

Fact #326: Peacocks Are Bad Luck Indoors

Peacock feathers are believed to bring bad luck if kept indoors because the "eye" pattern on the feathers is associated with the evil eye. Beautiful but risky décor!

Fact #327: Owls Are Believed to Steal Souls

In some Native American and African cultures, owls are thought to steal souls or bring messages from the afterlife. If an owl stares at you too long, don't take it personally—just walk away.

Fact #328: Spiders at Night Mean Trouble

In some cultures, seeing a spider at night is said to bring bad luck. By morning, however, they're considered lucky. So if you spot one after dark, maybe just pretend you didn't see it.

Fact #329: Your Ears Burning Means Someone's Talking About You

This old superstition suggests that if your ears are burning, someone is gossiping about you. The solution? Bite your tongue and make it stop. Sounds fair.

Fact #330: Eating Grapes at Midnight Brings Good Luck

In Spain, eating 12 grapes at midnight on New Year's Eve is thought to bring good luck for the year ahead—one grape for each month. Just don't choke on your good fortune.

Fact #331: A Bird in the House Is a Bad Omen

In many cultures, a bird flying into your home is believed to bring bad luck or symbolize death. Good thing most windows have screens these days.

Fact #332: Carrying an Acorn Prevents Aging

In ancient times, women carried acorns in their pockets because they believed it would keep them young. Maybe it's the original anti-aging hack?

Fact #333: Left-Handed People Were Once Feared

In the Middle Ages, being left-handed was considered a sign of witchcraft or evil. Thankfully, we've moved past this—now it's just a cool trait.

Fact #334: Saying "Rabbit" for Luck

In England, saying "rabbit" or "white rabbit" first thing in the morning on the first day of the month is believed to bring good luck. Sounds like a quirky way to start your day.

Fact #335: Butterflies Are Spirit Messengers

Many cultures believe butterflies are the spirits of deceased loved ones visiting the living. It's a comforting thought—unless you're afraid of bugs.

Fact #336: Never Give a Knife as a Gift

In many cultures, gifting a knife is thought to "cut" the relationship. The workaround? The recipient has to "pay" you with a small coin to cancel out the bad luck.

Fact #337: The Curse of the Umbrella Indoors

Not only does an open umbrella indoors supposedly bring bad luck, but in Victorian England, it was believed to also offend the sun gods. No wonder they loved their parasols instead.

Fact #338: Horseshoes Bring Good Luck (If Hung Properly)

A horseshoe is lucky if hung with the ends pointing up to hold the luck inside. Hung upside-down? The luck spills out, and nobody wants that!

Fact #339: Bells Drive Away Evil Spirits

In many cultures, ringing bells is believed to scare off evil spirits. That's why you often see bells in ceremonies or on Christmas decorations—double duty for festive vibes and ghost protection.

Fact #340: Sweeping Someone's Feet Brings Bad Luck

In Brazil, sweeping someone's feet with a broom is believed to curse them to never marry. Maybe just aim for the dirt next time.

Fact #341: Never Toast With Water

In Germany and some other countries, toasting with water is thought to wish death upon the people you're drinking with. Cheers… but only with wine or soda, please.

Fact #342: Moths Symbolize Death

In Mexico and some other cultures, seeing a black moth inside your home is a sign of death. Looks like butterflies' nocturnal cousins got the short end of the superstition stick.

Fact #343: An Itchy Palm Means Money

In some cultures, an itchy right palm means you'll receive money, while an itchy left palm means you'll lose it. Keep track of your palms—you might be onto something.

Fact #344: Wearing Red Ward Off Ghosts

In Chinese culture, wearing red is thought to protect against evil spirits. It's also why red is so prominent at weddings and festivals—it's the color of joy and safety.

Fact #345: A Broken Clock Stops Time (Sort Of)

A broken clock is believed to stop time in some cultures, symbolizing misfortune or even death. Better fix it or toss it— just don't let it tick people off.

Fact #346: Whistling in the Wind Brings Trouble

In parts of Asia, whistling on windy days is said to call spirits into your home. Better to just hum your favorite tune instead.

Fact #347: Never Turn Bread Upside Down

In Italy, placing bread upside down on a table is believed to bring bad luck. Bread is considered sacred, so treat it with respect—or at least keep it right-side up.

Fact #348: Blinking Lights Are a Spirit's Calling Card

In many cultures, flickering lights or sudden power outages are believed to be caused by ghosts or spirits trying to communicate. Or, you know, just bad wiring.

Fact #349: Hiccups Mean Someone's Talking About You

Some believe that hiccups mean someone is gossiping about you. No word on how to tell if it's good gossip or bad, though.

Fact #350: A Cat Washing Its Face Predicts Visitors

In Japan, if a cat washes its face with its paw, it's a sign that guests will arrive soon. Cats: adorable furballs and future-tellers.

Fact #351: Never Open Two Doors at the Same Time

In some cultures, opening two doors at the same time is said to create a "pathway" for spirits to enter. Who knew doors could be so supernatural?

Fact #352: Avoid the Number Four in East Asia

In countries like China and Japan, the number four is considered unlucky because it sounds like the word for "death." That's why some buildings skip the fourth floor entirely—just like we do with 13.

Fact #353: Knives Under the Pillow Ward Off Nightmares

In parts of Italy and Greece, placing a knife under your pillow is thought to protect you from bad dreams. A little extreme, but who doesn't want to sleep in peace?

Fact #354: Never Gift a Clock

In Chinese culture, gifting someone a clock is considered bad luck because it symbolizes the passage of time and impending death. Stick to gift cards instead!

Fact #355: Don't Sweep After Sundown

In some Indian households, sweeping at night is believed to sweep away good luck and wealth. Better leave the cleaning for daylight hours.

Fact #356: Spilling Water Behind Someone Brings Good Luck

In Serbia, spilling water behind someone as they leave the house is thought to bring them good fortune on their journey. It's like saying, "Break a leg," but wetter.

Fact #357: Birds Flying Into Windows Are Bad News

A bird crashing into your window is believed to bring bad luck or even death in some cultures. Turns out, window decals are lifesavers for birds and your nerves.

Fact #358: Horseshoes Over the Door Keep Spirits Out

Hanging a horseshoe above your front door is thought to ward off evil spirits. Just make sure the open end faces up to keep the luck from "spilling out."

Fact #359: A Dog Howling at Night Means Trouble

In many cultures, a dog howling in the middle of the night is considered an omen of death. Let's just hope they're howling at the moon instead.

Fact #360: White Moths Are Visitors from the Dead

In Filipino folklore, white moths are believed to be the spirits of deceased loved ones visiting the living. Who knew moths could be so sentimental?

Fact #361: Chewing Gum at Night Attracts Ghosts

In Turkey, it's said that chewing gum after dark turns it into the flesh of the dead. Honestly, this might just be a sneaky way to discourage late-night snacking.

Fact #362: Don't Look Back When Leaving a Funeral

In some cultures, looking back while leaving a funeral is believed to invite the spirit of the deceased to follow you home. Keep your eyes forward and your ghosts elsewhere!

Fact #363: Putting a Hat on a Bed Brings Bad Luck

In some regions, placing a hat on a bed is thought to bring misfortune. No one's quite sure why, but better safe than sorry—just use a hat rack.

Fact #364: Cutting Your Hair on a Full Moon Brings Prosperity

Some believe that trimming your hair during a full moon can make it grow faster and healthier. Haircuts and moon magic—what a combo!

Fact #365: Always Enter and Leave Through the Same Door

In some cultures, entering and exiting through different doors is thought to bring bad luck. Pick a door and stick to it—it's not that hard.

Fact #366: A Butterfly Landing on You Brings Good Luck

In many cultures, a butterfly landing on you is a sign of upcoming good fortune. Unless you're scared of bugs, this one's a win.

Fact #367: Don't Cross Knives on the Table

Crossed knives on a table are believed to invite conflict or bad luck. It's probably just a way to avoid awkward mealtime fights.

Fact #368: Hearing Footsteps Behind You is a Sign of a Ghost

In some cultures, unexplained footsteps behind you at night are thought to belong to a spirit. Time to pick up the pace and not look back.

Fact #369: Odd Numbers Are Luckier Than Even Numbers

In some cultures, odd numbers are considered more auspicious because they can't be easily divided, symbolizing unity and prosperity.

Fact #370: A Broken Eggshell Lets Evil Spirits Escape

In parts of Europe, it's believed that leaving broken eggshells lying around lets evil spirits hide inside. Always crush them after breakfast—just in case.

Fact #371: Whistling in the Car is a No-Go

In Lithuania, whistling in a car is said to summon evil spirits or bad luck. Maybe just sing along to the radio instead.

Fact #372: Cats Can Steal Your Soul While You Sleep

Some old superstitions suggest cats might sit on your chest while you sleep and steal your soul. Turns out, they're probably just looking for a warm spot to nap.

Fact #373: Dropping Silverware Predicts Visitors

In the Philippines, there's a belief that dropping a utensil signals the arrival of a guest. A dropped fork is thought to indicate a male visitor, while a spoon suggests a female visitor is coming.

Fact #374: A Chilly Breeze Means a Spirit Is Near

Feeling a sudden cold draft in a warm room? In many cultures, it's thought to be the presence of a ghost. Or maybe just a drafty window.

Fact #375: Open Scissors Bring Bad Luck

Leaving scissors open is believed to invite bad luck—or even malevolent spirits. Plus, it's just dangerous, so better to keep them closed.

Fact #376: Don't Sleep With Mirrors Facing Your Bed

In feng shui, a mirror facing your bed is thought to invite spirits into your dreams. Time to rethink your bedroom layout.

Fact #377: Picking Up a Penny Brings Good Luck—But Only If It's Heads Up

Finding a penny heads-up is believed to bring good luck, while tails-up is considered bad luck. If it's tails, the rule is to leave it where it lies for someone else to find their fortune. Even coins have their own superstitions!

Fact #378: Don't Sleep During Twilight Hours

In some cultures, sleeping during twilight (the time between day and night) is believed to bring nightmares or bad luck. Time your naps carefully!

Fact #379: Knives Are Never Given at Weddings

In many cultures, giving knives as wedding gifts is considered bad luck because they symbolize cutting the relationship. Better stick to blenders and gift cards.

Fact #380: An Itchy Nose Means a Fight Is Coming

If your nose starts itching, some superstitions say it's a sign you're about to get into an argument. Time to take a deep breath and avoid any drama.

Fact #381: Broken Plates Bring Good Luck at Weddings

In Greece, smashing plates at weddings is thought to bring good fortune and ward off evil spirits. Plus, it's a great excuse to make a mess.

Fact #382: Whistling Past a Cemetery is Forbidden

In some cultures, whistling while passing a cemetery is believed to wake the dead or disturb spirits. Better to stay quiet and respectful.

Fact #383: Opening a Window After Someone Dies Releases Their Spirit

Many cultures believe that opening a window immediately after someone passes allows their spirit to leave peacefully. It's a simple but comforting tradition.

Fact #384: Hanging Garlic Keeps Vampires Away

The idea that garlic wards off vampires comes from folklore, where it was thought to repel evil spirits. If nothing else, it keeps your breath strong enough to scare anyone away.

Fact #385: Crossing Paths with a Funeral Procession is Bad Luck

In some traditions, encountering a funeral procession while walking is said to bring bad fortune. It's best to let it pass and step aside.

Fact #386: Don't Cut a Baby's Hair Before Their First Birthday

In many cultures, cutting a baby's hair before their first birthday is thought to bring bad luck. It's also a handy way to avoid early haircut drama.

Fact #387: Sweeping Dust Toward the Door Sweeps Away Wealth

In Chinese superstition, sweeping dust or dirt toward the front door is believed to push out good fortune. Sweep inward instead to keep the luck inside!

Fact #388: Mirrors Should Be Covered During Thunderstorms

Some old superstitions warn that mirrors can attract lightning during a storm. Whether true or not, it's a great excuse to avoid looking at yourself on a rainy day.

Fact #389: Spiders in Your Home Mean Good Luck

While many people fear spiders, some superstitions say they bring good fortune if found indoors. Just maybe not in your shoes, please.

Fact #390: Don't Point at the Moon

In some Asian cultures, pointing at the moon is thought to anger the moon goddess, who might punish you with a cut or sore on your finger. Better just admire it from afar.

Fact #391: Crickets Are Good Luck

In many cultures, hearing a cricket in your home is a sign of good luck. So if one sneaks in, maybe think twice before shooing it out.

Fact #392: Odd Numbers of Flowers Are Lucky

In Russia, giving an odd number of flowers is seen as good luck, while an even number is reserved for funerals. Florists there know how to keep their bouquets balanced.

Fact #393: Never Turn Your Bed During a Storm

Some believe that moving your bed during a thunderstorm invites bad luck or restless spirits. Better to wait until the weather clears before redecorating.

Fact #394: Eating Fish on New Year's Day Brings Prosperity

In many cultures, fish is considered a lucky food to eat on New Year's Day because its scales symbolize wealth, and fish swim forward—just like you should in the new year.

Fact #395: Crossing Fingers Isn't Always for Luck

In some cultures, crossing your fingers behind your back is seen as a sign of lying or bad faith. Use this gesture wisely, or better yet, not at all.

Fact #396: Opening an Umbrella in the House Brings Arguments

In addition to bad luck, some superstitions suggest opening an umbrella indoors invites conflict and quarrels. Best to keep it closed until you're outside.

Fact #397: The First Visitor on New Year's Day Determines Luck

In Scotland, the "first-foot" tradition says the first person to enter your home on New Year's Day brings good or bad luck for the year ahead. Bonus points if they bring a gift like coal or bread.

Fact #398: Never Hand Someone a Knife Directly

In some cultures, passing a knife directly to someone is thought to cut the bond of friendship. To avoid bad luck, place the knife on a surface for them to pick up instead.

Fact #399: Lighting Three Cigarettes on One Match Brings Misfortune

This superstition stems from wartime, where lighting three cigarettes on one match gave snipers enough time to aim. It's now considered bad luck—or just wasteful.

Fact #400: Never Leave a Rocking Chair Empty

In Irish folklore, an empty rocking chair is an invitation for spirits to sit in it. If it starts moving on its own? Time to find a new house!

POP QUIZ

Think you've got what it takes to conquer the world of strange and spooky superstitions? Let's test your luck (and your memory) with these 10 fun questions!

Questions

1. **True or False:** Knocking on wood originated from the belief that spirits lived in trees.
2. What number is considered unlucky in many East Asian cultures because it sounds like the word for "death"?
3. **True or False:** It's believed that placing a hat on a bed brings good luck.
4. What everyday kitchen item is said to ward off vampires in folklore?
5. **True or False:** Spiders found indoors are considered lucky in some cultures.
6. According to Turkish beliefs, who is said to be summoned by whistling at night?
7. What item is commonly hung above doorways to ward off evil spirits, as long as it's hung upright?
8. **True or False:** Chewing gum at night in Turkey is believed to turn it into the flesh of the dead.
9. What unusual tradition in Scotland involves the "first visitor" to your home on New Year's Day?
10. **True or False:** A butterfly landing on you is considered a sign of bad luck.

Answers

1. **True**: Knocking on wood originated from ancient beliefs that spirits lived in trees and could bring good luck or protection.
2. **The number 4**: In China and Japan, the number four is considered unlucky because it sounds like the word for "death" in their languages.
3. **False**: Placing a hat on a bed is actually considered bad luck in many cultures. Time to invest in a hat rack!
4. **Garlic**: Folklore says garlic is a natural vampire repellent— and it's also handy for cooking delicious meals.
5. **True**: Spiders are considered lucky in some cultures, symbolizing prosperity and good fortune. Just don't let them crawl on you!
6. **The devil**: In Turkey, whistling at night is believed to summon the devil.
7. **Horseshoe**: Hanging a horseshoe above a doorway is believed to ward off evil spirits, as long as it's hung with the open end facing up to "hold" the luck.
8. **True**: In Turkey, chewing gum at night is thought to turn into the flesh of the dead. Creepy—and definitely a good excuse to skip late-night gum.
9. **First-footing**: In Scotland, the first person to enter your home on New Year's Day is thought to determine your luck for the year. A gift like bread or coal boosts the good vibes!
10. **False**: A butterfly landing on you is considered a sign of good luck, often symbolizing transformation or a visit from a loved one's spirit.

How did you do? Whether you aced it or learned something new, superstitions always keep life interesting!

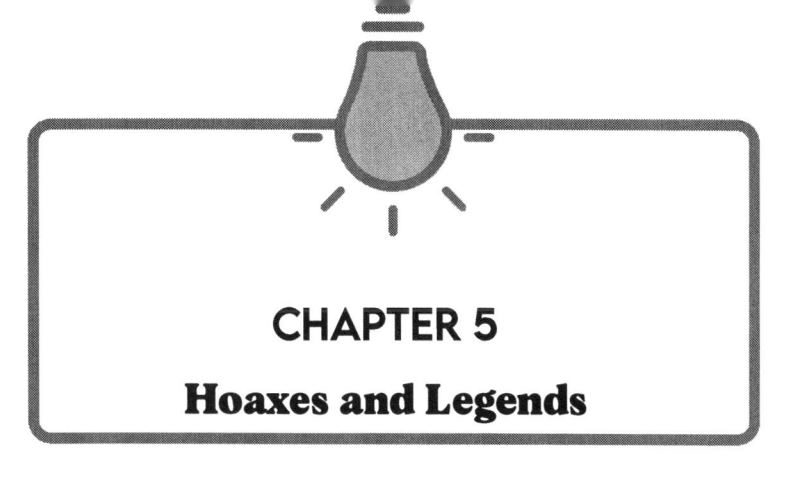

CHAPTER 5

Hoaxes and Legends

The world is full of wild hoaxes and legends that have tricked, entertained, and baffled people for centuries. From fake monsters to hilariously clever pranks, these stories blur the line between reality and imagination. Buckle up for a ride through the weirdest tales that had everyone scratching their heads (or laughing out loud)!

Fact #401: Bigfoot Might Be the Best at Hide-and-Seek

Bigfoot, the legendary ape-like creature, has been "spotted" thousands of times, yet no one has managed to catch one. Either Bigfoot is a master of camouflage, or people just need better cameras.

Fact #402: The Loch Ness Monster's Most Famous Photo Was a Prank

The iconic photo of the Loch Ness Monster, known as the "Surgeon's Photograph," turned out to be a toy submarine with a fake head. Even Nessie must've laughed at how long people believed it.

Fact #403: Crop Circles Were Made by Two Guys With Planks

For years, people thought aliens were behind mysterious crop circles, but two pranksters from England admitted they made most of them using wooden planks. Turns out, extraterrestrials have a great sense of humor—or none at all.

Fact #404: The Cardiff Giant Was a Gigantic Lie

In 1869, workers in New York "discovered" a petrified 10-foot-tall man. It turned out to be a hoax by George Hull, who carved the giant from stone and buried it as a prank. People still paid to see it—because why not?

Fact #405: The Cottingley Fairies Fooled Everyone, Even Sherlock Holmes

Two young girls in 1917 took photos of what they claimed were real fairies. The pictures fooled many, including Sir Arthur Conan Doyle, creator of Sherlock Holmes. Decades later, the girls admitted they used paper cutouts. Even the sharpest detective wouldn't have guessed!

Fact #406: P.T. Barnum Once Displayed a Fake Mermaid

The famous showman P.T. Barnum exhibited the "Fiji Mermaid," a bizarre creation made by sewing a monkey's head onto a fish's body. People paid to see it, proving that curiosity really is priceless—and weird.

Fact #407: A Man Convinced the World He Found a UFO

In 1957, a man in Brazil claimed to have found pieces of a flying saucer. Scientists later discovered it was made of materials from Earth, but it still got people buzzing about extraterrestrial life.

Fact #408: Orson Welles Scared America With a Fake Alien Invasion

In 1938, Orson Welles' radio play *War of the Worlds* caused mass panic as listeners believed Martians were actually invading Earth. It was just a fictional story—imagine how they'd react to a Hollywood blockbuster!

Fact #409: The Jersey Devil Has Been Spotted for Over 200 Years

The Jersey Devil, a legendary creature said to haunt the Pine Barrens of New Jersey, has terrified locals since the 1700s. It's described as having wings, hooves, and a piercing scream. Sounds like a bad mood personified.

Fact #410: The Piltdown Man Hoax Fooled Scientists for Decades

In 1912, archaeologists in England claimed to have found the "missing link" between apes and humans. It took over 40 years to discover it was a skull made of human and orangutan bones glued together. That's one sticky lie!

Fact #411: Jackalope: A Rabbit With Antlers?

The jackalope, a mythical creature with rabbit ears and antelope antlers, originated as a joke by two Wyoming hunters. Today, it's a pop culture icon—proof that fake animals can be just as fun as real ones.

Fact #412: Spaghetti Grows on Trees (According to the BBC)

In 1957, the BBC aired a prank segment showing spaghetti being harvested from trees in Switzerland. Viewers actually called in asking how to grow their own spaghetti trees. Now that's a recipe for laughs!

Fact #413: The Moon Landing Was NOT a Hoax (But Some People Still Believe It)

Despite overwhelming evidence, some conspiracy theorists insist the 1969 moon landing was faked on a Hollywood set. Neil Armstrong must be rolling his eyes in space.

Fact #414: A "Sea Monster" Was Just a Whale Carcass

Many "sea monster" sightings turn out to be decomposing whale carcasses. The ocean is mysterious, but sometimes it's just gross.

Fact #415: The Tasmanian Tiger Is Still Hiding... Or Is It?

Although declared extinct in 1936, the Tasmanian tiger is still reported in sightings today. It's like the Bigfoot of the animal world—if Bigfoot had stripes.

Fact #416: Cottingley Fairies Were Inspired by a Comic

The paper cutouts used by the Cottingley girls to fake fairy photos were based on illustrations from a children's comic. Even hoaxes need artistic inspiration.

Fact #417: The Great Moon Hoax of 1835

A New York newspaper published stories claiming astronomers found life on the moon, including unicorns and bat-people. Readers ate it up—because who wouldn't want lunar unicorns?

Fact #418: Nessie Might Be a Giant Eel

Some scientists speculate the Loch Ness Monster could be a giant eel, which would explain its long, slippery sightings. Nessie's PR team must hate this theory.

Fact #419: The Phantom Time Hypothesis Claims We Skipped 297 Years

This wild conspiracy theory suggests that three centuries of history (614–911 AD) never happened and were invented by medieval rulers. So technically, it's not 2024—it's 1727. Who wants to check the math?

Fact #420: Ghost Photos Are Often Camera Tricks

Many famous ghost photos turn out to be reflections, double exposures, or dust on the lens. The ghosts? Just photobombing accidents.

Fact #421: The Treasure That Never Was

Legend has it that Japanese soldiers hid a massive treasure in the Philippines during World War II. Despite decades of searching, no evidence of "Yamashita's Gold" has ever been found, making it one of the Philippines' most enduring hoaxes.

Fact #422: A Fake Yeti Scalp Fooled Explorers

A monastery in Nepal displayed what it claimed was a Yeti scalp. When tested, it turned out to be made from goat skin. Even Yetis have bad hair days.

Fact #423: The Bermuda Triangle Is Just Bad Weather and Math

Many Bermuda Triangle disappearances can be explained by storms, magnetic interference, and human error. Sorry, no aliens or portals—just bad luck.

Fact #424: A 19th-Century Vampire Panic Swept New England

In the 1800s, some families dug up graves to "kill" suspected vampires, believing they caused mysterious illnesses. Turns out, it was probably tuberculosis.

Fact #425: The Montauk Monster Was Just a Raccoon

The Montauk Monster, a strange creature washed ashore in New York, sparked theories of government experiments. Scientists later confirmed it was a decayed raccoon. Mystery solved—sort of.

Fact #426: The Mokele-Mbembe Might Be Africa's Loch Ness Monster

Deep in the Congo River, locals claim to have seen a creature resembling a dinosaur. Scientists haven't found proof, but the legend of the Mokele-Mbembe lives on.

Fact #427: Paul McCartney's Death Hoax

In 1969, a bizarre rumor spread that Paul McCartney had died and was replaced by a lookalike. Fans even "found clues" in Beatles songs, like "I buried Paul" in *Strawberry Fields Forever*. Spoiler: Paul's alive and well.

Fact #428: The Goatman Terrorized Maryland

The Goatman, a half-man, half-goat cryptid, is rumored to roam the woods of Maryland, scaring hikers and campers. Whether he's real or not, he's definitely a terrible roommate for nature lovers.

Fact #429: The Hoax of the Cottingley Goblins

Inspired by the famous Cottingley Fairies, a prankster in the 1940s claimed to have photographed "goblins" in a woodland area in England. Though the photos were debunked as staged, they sparked a brief cultural obsession with mythical creatures lurking in the countryside.

Fact #430: A Fake Time Traveler Fooled the Internet

In the early 2000s, a man named John Titor claimed to be a time traveler from the year 2036. He shared wild predictions, none of which came true. Nice try, John!

Fact #431: The Winchester Mystery House Was Designed to Confuse Ghosts

The Winchester Mystery House, with its maze-like layout, was built by Sarah Winchester to confuse the spirits she believed haunted her family. Ghost-proofing your home: a Victorian trend.

Fact #432: The Hope Diamond Is Cursed

This famous diamond is said to bring misfortune to its owners. Stories of bad luck include deaths, financial ruin, and general chaos. But hey, it still looks fabulous in a museum.

Fact #433: The "Dead Fairy" Hoax Fooled the Internet

A man in the UK claimed to find the mummified remains of a fairy, complete with tiny wings. He later admitted it was a prop, but not before it went viral. The internet never forgets.

Fact #434: A 100-Year-Old Map Sparked Atlantis Rumors

In 1665, a map appeared to show Atlantis, the mythical sunken city. Scholars later debunked it as a misinterpretation, but the legend of Atlantis continues to make waves.

Fact #435: Spring-Heeled Jack Terrorized Victorian London

This bizarre urban legend describes a figure in London who could jump great heights and breathe fire. Some say he was a prankster, while others swear he was supernatural.

Fact #436: The Phantom Clown Scares

In the 1980s, reports of creepy clowns trying to lure children surfaced in the U.S., but no evidence was ever found. Sounds like the plot of a horror movie—and a terrible career move for clowns.

Fact #437: The Alien Autopsy Hoax Fooled Millions

In 1995, a video claiming to show a government autopsy of an alien from the 1947 Roswell UFO incident was released. The footage sparked worldwide intrigue but was later revealed to be staged using props and a mannequin. Even decades later, the hoax remains one of the most famous in UFO lore.

Fact #438: The Jersey Devil Was Blamed for Livestock Deaths

In the early 1900s, farmers blamed the Jersey Devil for killing livestock. Whether it's real or not, it's an excellent scapegoat—literally.

Fact #439: A "Tree Monster" Terrified a Town

In 2017, residents in Poland reported a strange creature in a tree. It turned out to be a croissant. A delicious legend in the making.

Fact #440: The Ropen: A Flying Cryptid in Papua New Guinea

Locals in Papua New Guinea claim to see a glowing, dinosaur-like creature called the Ropen. Scientists suspect it could be a large bat, but the jury's still out.

Fact #441: The Hoax of the Merman of Banff

In the early 1900s, a supposed "merman" was displayed in Banff, Canada. This creepy creature, said to be half-man and half-fish, was later revealed as a taxidermy creation using a monkey's torso and a fish's tail. Visitors were both terrified and fascinated by this macabre fabrication.

Fact #442: An Octopus in a Tree?

In Washington state, people once claimed to see an octopus in a tree. It was later revealed to be an elaborate art installation. Still, a tree-dwelling octopus would be pretty cool.

Fact #443: The Slenderman Hoax Got Out of Control

Slenderman, a fictional character created for an internet contest, became so famous that people started believing he was real. Proof that the internet has its own cryptids.

Fact #444: The Curse of the Koh-i-Noor Diamond

The Koh-i-Noor diamond, part of the British Crown Jewels, is rumored to bring misfortune and death to any man who possesses it. The diamond has a bloody history, having been passed between rulers in battles, often resulting in their untimely demise. Women, however, are believed to be immune to its curse.

Fact #445: The Great Airship Mystery of 1897

Reports of mysterious airships appeared across the U.S. decades before airplanes were invented. Were they alien crafts—or just creative tall tales?

Fact #446: A Prankster "Invented" a New Animal

In the 1700s, naturalist George Steller described the Steller's Sea Ape—a creature he later admitted to making up just to confuse people. Well played, George.

Fact #447: The Beast of Bodmin Moor

In England, sightings of a large, black panther-like creature on Bodmin Moor have baffled locals for years. Scientists suggest it could be a big cat escaped from captivity—or just a really scary house cat.

Fact #448: An Island Disappeared from Maps

Sandy Island appeared on maps for over a century but doesn't actually exist. It was officially "undiscovered" in 2012. Talk about a cartographer's oops moment.

Fact #449: A Phantom Army Was Seen in WWII

During World War II, soldiers in Normandy reported seeing ghostly armies marching across the battlefield. Whether it was stress or spirits, the sightings remain unexplained.

Fact #450: Anastasia's Survival Hoax

One of the most famous myths claims that Anastasia Romanov, the youngest daughter of Tsar Nicholas II, survived the execution of the royal family in 1918. Over the years, multiple women claimed to be Anastasia, including Anna Anderson, but DNA evidence has since confirmed the entire family perished.

Fact #451: The Green Children of Woolpit

In 12th-century England, two children with green skin reportedly appeared in a village. They spoke an unknown language and claimed to come from a land called St. Martin. The story remains an unsolved mystery—or a medieval prank.

Fact #452: The Vanishing Village of Angikuni Lake

In 1930, a Canadian fur trapper claimed that an entire Inuit village disappeared without a trace. Though later debunked, the tale of the vanishing village still sends shivers down spines.

Fact #453: The Monster of Lake Champlain

Known as "Champ," this cryptid is said to live in New York's Lake Champlain. Sightings go back centuries, but so far, no one's caught Champ on a fishing line. Smart monster.

Fact #454: The Great Balloon Hoax

In 1844, a newspaper claimed a man crossed the Atlantic in a hot air balloon. It was completely fake, but the story got people dreaming of transatlantic travel—decades before planes made it real.

Fact #455: The Marcos Golden Buddha

A story circulated that Ferdinand Marcos found a golden Buddha statue filled with diamonds as part of Yamashita's Treasure. The alleged finder later sued the family and won, but the treasure was never proven to exist.

Fact #456: The Beast of Gévaudan

In 18th-century France, a wolf-like creature reportedly terrorized villagers, killing over 100 people. Some believe it was a giant wolf, while others think it was an escaped lion. Either way, no one volunteered to investigate up close.

Fact #457: The Alien Lights of Marfa, Texas

For over a century, mysterious lights have appeared in the skies above Marfa, Texas. Theories range from UFOs to car headlights, but no one knows for sure what they are. Texas keeps its secrets.

Fact #458: The Great Wall of China Is NOT Visible from Space

One of the most persistent hoaxes is the claim that the Great Wall of China is visible from the Moon. In reality, astronauts have confirmed that it's nearly impossible to see with the naked eye, even from low Earth orbit. Space doesn't lie— tourism ads sometimes do.

Fact #459: The Hoax That Launched Shark Week

In 2004, Discovery Channel aired a mockumentary about a 38-foot shark named *Megalodon*. It was fake, but viewers took it seriously—and it made Shark Week a massive hit.

Fact #460: The Spooky Legend of the Black Eyed Children

Urban legends describe eerie children with completely black eyes knocking on doors and asking to come inside. The rule? Never let them in—or you might regret it.

Fact #461: The Mystery of the Taos Hum

Residents of Taos, New Mexico, report hearing a low, droning hum with no identifiable source. Scientists have tried to find an explanation, but the hum remains a bizarre auditory mystery.

Fact #462: The Lizard Man of Scape Ore Swamp

In South Carolina, sightings of a half-man, half-lizard creature have been reported since the 1980s. Whether it's a cryptid or a guy in a really good costume, locals still share the story.

Fact #463: A Ghost Ship Was Found in 1872

The Mary Celeste, a ship found adrift with no crew aboard, has puzzled investigators for over a century. Was it pirates, mutiny, or something supernatural? No one knows.

Fact #464: The "Warminster Thing" UFO Sightings

In 1965, residents of Warminster, England, reported strange noises and glowing lights in the sky. Theories ranged from UFOs to secret military experiments, but the truth remains elusive.

Fact #465: The Hairy Hands of Dartmoor

Drivers in Dartmoor, England, claim ghostly hairy hands have grabbed their steering wheels, causing accidents. Whether it's a hoax or just bad driving, it's definitely nightmare fuel.

Fact #466: The Devil's Tramping Ground

In North Carolina, a patch of land where nothing grows is believed to be where the Devil paces at night. Scientists suggest soil issues, but the creepy vibe remains.

Fact #467: The Hoax That Fooled NASA

In 1976, a man named Gary Dahl invented "pet rocks" as a joke—and sold millions of them. While not NASA-related, it's proof people love quirky ideas.

Fact #468: The Alien Abduction of Barney and Betty Hill

In 1961, the Hills claimed they were abducted by aliens in New Hampshire. Their detailed story became one of the most famous UFO cases ever—and inspired countless sci-fi tales.

Fact #469: Imelda's 3,000 Pairs of Shoes

While Imelda Marcos did own a large shoe collection, the claim that she owned 3,000 pairs has been debunked, with the actual number closer to 1,000. Still, it became a symbol of excess during the dictatorship.

Fact #470: The Krampus Legend

Krampus, the Christmas demon, is said to punish naughty children in Alpine folklore. A terrifying alternative to Santa, Krampus even has his own parade. Who's ready to be on the nice list now?

Fact #471: The Disappearing Lighthouse Keepers

In 1900, three lighthouse keepers vanished from their post on Scotland's Flannan Isles. Theories range from rogue waves to alien abduction, but their fate remains unknown.

Fact #472: A UFO Battle Over Nuremberg

In 1561, residents of Nuremberg, Germany, reported seeing strange flying objects in the sky, which they described as a "battle." It's one of the earliest recorded UFO sightings—and still a mystery.

Fact #473: The Vanishing Hitchhiker Legend

This urban legend tells of ghostly hitchhikers who disappear from the car after being picked up. Whether fact or fiction, it's enough to make you think twice about roadside strangers.

Fact #474: Hitler's Diary Was a Forgery

In 1983, a German magazine paid millions for what they thought were Adolf Hitler's personal diaries. It turned out they were clumsy fakes written with modern materials. The scandal was as shocking as the content was fake.

Fact #475: The Mysterious Moa Sightings

The moa, a giant bird thought to be extinct, is occasionally reported in New Zealand. Are they hiding—or just really good at playing extinct?

Fact #476: The Ghost of Flight 401

After Flight 401 crashed in 1972, parts of the plane were salvaged and used in other aircraft. Passengers and crew on those planes later reported seeing ghostly figures of the deceased flight crew, including the captain. These sightings spooked everyone so much that the airline eventually removed the salvaged parts!

Fact #477: The Beast of Bray Road

In Wisconsin, sightings of a werewolf-like creature along Bray Road date back to the 1930s. Locals describe it as part-wolf, part-human, and all terrifying. Full moons just got a lot scarier.

Fact #478: The Phantom Dogs of Black Shuck

In England, tales of ghostly black dogs with glowing red eyes have haunted coastal regions for centuries. Known as Black Shuck, these spectral canines are said to bring bad luck—or worse.

Fact #479: The Loveland Frogman

Ohio has its own cryptid: the Loveland Frogman, a humanoid frog reported near the Little Miami River. Whether it's an escaped science experiment or a prank, this frogman has leaped into legend.

Fact #480: The Curse of Ötzi the Iceman

Ötzi, a 5,300-year-old mummy discovered in the Alps, is rumored to curse those who disturb him. Several researchers involved in his study have died under strange circumstances. Coincidence? Maybe.

Fact #481: The Tunguska Event Was Likely a Meteor

In 1908, a massive explosion flattened 800 square miles of Siberian forest. Many believed it was a UFO crash, but scientists think it was a meteor or comet. Either way, it was cosmic chaos.

Fact #482: The Green Flash at Sunset

A rare optical phenomenon, the green flash occurs just as the sun sets or rises. Sailors once thought it was a sign of magic or hidden treasure. Turns out, it's just science showing off.

Fact #483: The Ghostly Bells of Whaley House

The Whaley House in California is considered one of the most haunted houses in the U.S. Visitors have reported hearing disembodied footsteps and ghostly bells. Maybe the ghosts are just welcoming guests.

Fact #484: The El Chupacabra Mystery

Reports of a blood-sucking creature called El Chupacabra first appeared in Puerto Rico in the 1990s. The legend spread worldwide, with farmers blaming it for livestock deaths. Alien, cryptid, or just a really angry coyote? No one knows.

Fact #485: The Fresno Nightcrawlers

Videos from Fresno, California, show strange, walking stick-like creatures dubbed "nightcrawlers." The footage has baffled experts—and terrified anyone who's watched it late at night.

Fact #486: The Mystery of Area 51

Area 51 has been at the center of UFO and alien conspiracy theories for decades. While the U.S. government admits it exists, its real purpose remains classified. Cue the alien memes.

Fact #487: The Black Knight Satellite Conspiracy

Some believe a mysterious satellite called the Black Knight has orbited Earth for 13,000 years, sent by extraterrestrials. NASA says it's just space debris, but conspiracy theorists aren't convinced.

Fact #488: The Yeti's Cousin, the Almas

In Mongolia, legends of the Almas—a humanoid, ape-like creature—persist. Unlike the elusive Yeti, the Almas is said to resemble a caveman. Cryptid or living fossil? The mystery continues.

Fact #489: The Devil's Bridge

Legends across Europe tell of bridges said to be built by the Devil in exchange for a soul. The eerie designs of these bridges only add to their spooky reputations.

Fact #490: The Sea Serpent of Gloucester

In the 1800s, New Englanders reported sightings of a 100-foot-long sea serpent off the coast. Despite multiple accounts, no physical evidence was ever found. Maybe it was just a giant eel on vacation.

Fact #491: The Vampire of Highgate Cemetery

In the 1970s, rumors of a vampire haunting Highgate Cemetery in London sparked vampire-hunting expeditions. While no vampire was ever found, the cemetery's creepy vibe remains legendary.

Fact #492: The Cursed Chair of Thomas Busby

Thomas Busby, a murderer in 1702, cursed his chair before being executed. Since then, several people who sat in the chair reportedly met untimely deaths. It now hangs in a museum—far out of reach.

Fact #493: The Riddle of the Georgia Guidestones

This mysterious monument in Georgia features cryptic inscriptions in multiple languages about humanity's future. No one knows who built it, and its message has sparked countless conspiracy theories.

Fact #494: The Moaning Myrtle Ghost

In the 1970s, construction workers in Scotland reported hearing ghostly moans while renovating a school. Locals believe it's haunted by a student who drowned there decades earlier.

Fact #495: The Phoenix Lights UFO Sighting

In 1997, thousands of people in Phoenix, Arizona, saw mysterious lights hovering in the sky. The military said it was flares, but UFO believers still insist it was extraterrestrial.

Fact #496: The Kraken Might Be Real

While the Kraken was long thought to be a myth, sightings of giant squids have fueled theories that the legendary sea monster has some truth behind it. Who's up for a deep-sea dive?

Fact #497: The Van Meter Visitor

In 1903, residents of Van Meter, Iowa, reported seeing a bat-like creature with a glowing horn. They tried to shoot it, but it disappeared into a mine. Sounds like the start of a superhero origin story.

Fact #498: The Devil's Sea

Off the coast of Japan, the Devil's Sea is rumored to have strange disappearances similar to the Bermuda Triangle. Sailors avoid it, and researchers are still trying to explain its mysterious occurrences.

Fact #499: The Hollow Earth Theory

Some conspiracy theorists believe the Earth is hollow and home to advanced civilizations or creatures. Spoiler: Science says it's solid—but the idea keeps popping up in sci-fi stories.

Fact #500: The Curse of the Poltergeist Movies

The *Poltergeist* film series is said to be cursed, with multiple cast members dying under mysterious circumstances. Whether it's coincidence or curse, it's enough to give anyone goosebumps.

POP QUIZ

Let's see how much you remember from this journey through mysterious hoaxes and baffling legends. Get ready for some brain-teasing fun!

Questions

1. **True or False:** The Loch Ness Monster's most famous photo was actually a toy submarine with a fake head.
2. What famous 19th-century hoax involved a supposedly petrified giant that was later revealed to be a carved gypsum statue?
3. **True or False:** P.T. Barnum displayed a "Fiji Mermaid" that was made of a monkey and a fish sewn together.
4. What 1938 radio broadcast caused mass panic because people thought it was reporting a real alien invasion?
5. **True or False:** The Mary Celeste was found adrift with no crew aboard, sparking ghost ship theories.
6. Which cursed diamond is rumored to bring misfortune to anyone who owns it?
7. **True or False:** The Cottingley Fairies photos were proven to be real.
8. What mysterious lights have been baffling residents in Marfa, Texas, for over a century?
9. **True or False:** The Winchester Mystery House was built to confuse ghosts.
10. What cryptid is said to haunt the Pine Barrens of New Jersey with wings, hooves, and a terrifying scream?

Answers

1. **True:** The famous Loch Ness Monster photo was a hoax, created with a toy submarine and fake head.
2. **Cardiff Giant:** A 10-foot "petrified man" carved as a prank still drew crowds, even after being exposed as fake.
3. **True:** P.T. Barnum's "Fiji Mermaid" was a bizarre mix of a monkey's head and a fish's body. Creepy, but irresistible to viewers.
4. **War of the Worlds:** Orson Welles' 1938 radio broadcast caused panic as listeners believed a Martian invasion was real.
5. **True:** The Mary Celeste was discovered adrift with its crew mysteriously missing. The case remains unsolved.
6. **Hope Diamond:** This legendary blue gem is said to curse its owners with misfortune and tragedy.
7. **False:** The Cottingley Fairies were just clever paper cutouts staged by two young girls.
8. **Marfa Lights:** Mysterious glowing orbs in Texas have baffled observers for years, inspiring UFO theories.
9. **True:** The Winchester Mystery House, with its maze-like design and staircases to nowhere, was built to confuse spirits.
10. **Jersey Devil:** A cryptid said to haunt New Jersey's Pine Barrens with eerie screeches and a terrifying appearance.

How did you score? Whether you aced it or learned some new legendary tales, the world of hoaxes and mysteries is full of surprises!

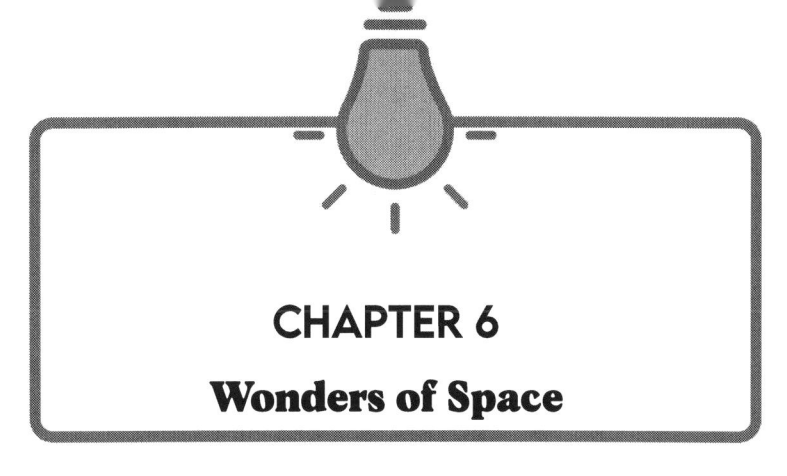

CHAPTER 6

Wonders of Space

Space is incredible, mysterious, and downright weird. It's a place where time bends, stars explode, and planets do things you'd only expect in sci-fi movies. Let's blast off into some hilarious, mind-blowing, and just plain bizarre cosmic facts!

Fact #501: Space Smells Like Barbecue

Astronauts have reported that the smell of space clings to their suits after a spacewalk, and it smells like... barbecue! Smoky, metallic, and a little burnt—like someone left the burgers on the grill too long.

Fact #502: A Day on Venus Lasts Longer Than Its Year

Venus takes 243 Earth days to spin once on its axis but only 225 days to orbit the Sun. Imagine living there: you'd have a birthday before your workday ended.

Fact #503: Saturn Could Float in a Bathtub

If you had a big enough bathtub (really big), Saturn would float because it's mostly made of gas. Good luck finding a rubber duck to match.

Fact #504: There's a Planet Made of Diamonds

A planet called 55 Cancri e is so rich in carbon that scientists believe it's covered in diamonds. Talk about a cosmic bling factory!

Fact #505: The Moon is Moving Away from Earth

Every year, the Moon drifts about 1.5 inches farther from Earth. Don't worry, it'll take billions of years to really ghost us, but it's a good excuse to say, "It's not you, it's the orbit."

Fact #506: You Could Cry in Space, But Your Tears Wouldn't Fall

In microgravity, tears don't stream down your face—they form little bubbles that stick to your eyes. So, crying in space is like getting a built-in water balloon fight.

Fact #507: There's a Giant Cloud of Alcohol in Space

Near the center of our galaxy, there's a massive cloud of ethanol, a type of alcohol. It's enough to make 400 trillion pints of beer, but don't pack your party hat—it's way too far away.

Fact #508: Neutron Stars Are So Dense They Defy Logic

A sugar-cube-sized amount of material from a neutron star would weigh about a billion tons on Earth. So, if someone tells you to "carry your weight," you can always say, "At least it's not neutron star weight."

Fact #509: A Year on Uranus is 84 Earth Years

On Uranus, you'd have to wait 84 Earth years to celebrate your next birthday. Cake and candles would definitely feel a little overdue.

Fact #510: Mars Has the Largest Volcano in the Solar System

Olympus Mons on Mars is about three times the height of Mount Everest. Mars isn't just the red planet—it's also the drama queen of the solar system.

Fact #511: Space is Completely Silent

No one can hear you scream in space because sound waves can't travel in a vacuum. It's basically the universe's way of saying, "Shhh."

Fact #512: The Coldest Place in the Universe is on Earth (Technically)

Scientists created a spot colder than space inside a lab at MIT. Space itself gets down to about -455°F, but this lab hit a record -459°F. Humans really do love breaking records.

Fact #513: Stars Don't Really Twinkle

The twinkling of stars is just Earth's atmosphere messing with the light. Stars themselves are steady—but don't tell kids that, or it'll ruin every lullaby.

Fact #514: Jupiter Has a Giant Red Spot That's a Storm

The Great Red Spot on Jupiter is a storm so big it could swallow Earth. It's been raging for over 300 years, making your local weather seem pretty tame.

Fact #515: There's a Star That's Basically a Disco Ball

A star called BPM 37093 is crystallized carbon—essentially a giant diamond in space. If stars threw parties, this one would definitely be the centerpiece.

Fact #516: The Sun Is Losing Weight

Every second, the Sun loses about 4 million tons of mass as it burns. Before you start panicking, this diet plan has been in place for billions of years and will keep going for billions more.

Fact #517: A Day on Mars is Almost the Same as a Day on Earth

Mars has a day that's 24 hours and 39 minutes long. Close enough to Earth time to make it feel familiar—minus the lack of breathable air.

Fact #518: The Earth's Core is as Hot as the Sun

Earth's core reaches temperatures of about 10,800°F, similar to the surface of the Sun. So technically, we're all living next to a molten hot tub.

Fact #519: There's a Planet That Rains Glass

On a planet called HD 189733b, it rains sideways shards of glass, thanks to 5,400 mph winds. Umbrellas? Not recommended.

Fact #520: Black Holes Can "Burp"

After devouring matter, black holes can release powerful bursts of energy. So basically, they're cosmic creatures with bad table manners.

Fact #521: The Milky Way is on a Collision Course

In about 4.5 billion years, the Milky Way will collide with the Andromeda galaxy. No need to pack—we've got plenty of time to prepare for our galactic road trip.

Fact #522: Astronauts Grow Taller in Space

In microgravity, the spine stretches, and astronauts can grow up to two inches taller. Too bad the effect disappears once they return to Earth.

Fact #523: The Largest Structure in the Universe is a Hole

The Boötes Void is an enormous empty region of space with very few galaxies. It's basically the cosmic equivalent of a ghost town.

Fact #524: There's a "Space Cemetery" for Satellites

When satellites reach the end of their lives, many are sent to the "Point Nemo," the most remote spot on Earth. It's the ultimate cosmic retirement home.

Fact #525: Space Junk is Out of Control

There are about 27,000 pieces of human-made debris orbiting Earth. Future astronauts might need trash pickers just to navigate the cosmos.

Fact #526: The Moon Causes Earth's Tides

The Moon's gravitational pull creates tides on Earth, making it nature's ultimate pool party planner. Without it, oceans would be a lot less wavy.

Fact #527: Mercury Has Ice Despite Being Close to the Sun

The poles of Mercury contain craters with permanent shadow, allowing ice to exist. Who knew the hottest planet could have such a cool secret?

Fact #528: The Universe is Expanding Faster Than Expected

The universe isn't just expanding—it's speeding up as it grows. It's like the cosmos hit the gas pedal, and we're all along for the ride.

Fact #529: Venus Spins Backwards

Most planets spin in the same direction as their orbit, but Venus spins the other way. Scientists think a massive collision flipped it around. Venus: the rebel of the solar system.

Fact #530: Mars Has Blue Sunsets

On Mars, dust in the atmosphere scatters sunlight differently, creating stunning blue sunsets. It's like a cosmic Instagram filter—Mars edition.

Fact #531: There's a Planet That Orbits Three Stars

The planet HD 131399Ab orbits three stars at once, meaning it would experience triple sunsets. Move over, Tatooine—this is the real Star Wars planet.

Fact #532: A Day on Jupiter is 10 Hours Long

Jupiter spins so fast that its day is just 10 hours, despite being the largest planet in the solar system. Imagine rushing through your morning coffee at Jupiter speed!

Fact #533: The Sun is Actually White

The Sun appears yellow due to Earth's atmosphere, but it's actually white. Space, you've been messing with our color perception all along.

Fact #534: Earth Has a Cosmic Twin

Kepler-452b, a planet located 1,400 light-years away, is considered Earth's "cousin" due to its similar size and potential for water. Too bad it's way out of commuting distance.

Fact #535: A Spacesuit Costs $12 Million

NASA's spacesuits cost a whopping $12 million each, with most of that price going into the backpack and life support system. Who knew survival could be so expensive?

Fact #536: The Milky Way Smells Like Rum and Tastes Like Raspberries

The center of our galaxy contains ethyl formate, the chemical responsible for the smell of rum and the taste of raspberries. Space truly is a cosmic cocktail.

Fact #537: The Largest Volcano in the Universe is on Mars

Olympus Mons, the massive Martian volcano, is about the size of Arizona and three times taller than Mount Everest. Mars isn't just red—it's extreme.

Fact #538: There's a Star That's 10 Billion Years Old

HD 140283, nicknamed the "Methuselah star," is one of the oldest stars ever discovered, clocking in at 10 billion years. Talk about a senior citizen of space.

Fact #539: There's an Exoplanet That Rains Iron

On the exoplanet WASP-76b, temperatures are so high that it rains molten iron. Definitely not the kind of weather for picnics.

Fact #540: A Rocket Launch Can Be Heard From 100 Miles Away

The sound of a rocket launch is so loud that it can be heard over 100 miles away. Good thing space doesn't have noise ordinances!

Fact #541: Uranus Spins on Its Side

Uranus is tilted so much that it essentially rolls around its orbit like a cosmic bowling ball. Scientists think a giant impact caused its unusual rotation.

Fact #542: The International Space Station is Moving Fast

The ISS orbits Earth at 17,500 mph, meaning astronauts see 16 sunrises and sunsets every day. Time flies when you're literally flying.

Fact #543: Black Holes Can Stretch You Like Spaghetti

If you fall into a black hole, its immense gravity will stretch you into a long, thin shape—a process scientists call "spaghettification." Deliciously terrifying.

Fact #544: There's a Rogue Planet Drifting Through Space

Rogue planets, like CFBDSIR 2149-0403, float through space without orbiting a star. They're the ultimate loners of the cosmos.

Fact #545: The Cold Spot in Space is a Mystery

A region in the cosmic microwave background is unusually cold, and scientists still can't explain it. Some even theorize it could be evidence of another universe. Multiverse, anyone?

Fact #546: A Chunk of Mars Fell to Earth

Meteorites from Mars occasionally land on Earth, allowing scientists to study the red planet without ever leaving home. Thanks, space delivery service!

Fact #547: Space Junk Could Cause a Chain Reaction

If enough space debris collides, it could create a domino effect known as the Kessler Syndrome, making low Earth orbit too dangerous for satellites. A cosmic cleanup crew might be in order.

Fact #548: Venus is the Hottest Planet

Despite being second from the Sun, Venus has a thick atmosphere that traps heat, making it the hottest planet in the solar system. Mercury, take notes!

Fact #549: There's a Cosmic Speed Limit

The speed of light—186,282 miles per second—is the fastest anything can travel. Einstein set the rules, and the universe follows them.

Fact #550: Earth is the Densest Planet

Earth might not be the biggest, but it's the densest planet in the solar system, thanks to its metallic core. Heavyweight champion of the planets!

Fact #551: The Sun Contains 99.8% of the Solar System's Mass

The Sun is so massive that it makes up nearly all of the mass in our solar system. The planets, moons, and asteroids are just cosmic crumbs.

Fact #552: Stars Can Be Born in "Nurseries"

Nebulas, also known as stellar nurseries, are where stars are born. These colorful clouds of gas and dust create some of the universe's most stunning sights.

Fact #553: You Can't Belch in Space

In microgravity, gas doesn't separate from liquid, so astronauts can't burp. Just imagine holding in a soda-fueled burp for an entire space mission!

Fact #554: There Are More Stars Than Grains of Sand on Earth

The universe contains an estimated 200 billion trillion stars, far outnumbering all the grains of sand on Earth's beaches. Talk about cosmic overload.

Fact #555: The Moon Has Moonquakes

Just like Earth has earthquakes, the Moon has moonquakes. These are caused by tidal stresses from Earth's gravity. Even the Moon can't catch a break.

Fact #556: A Year on Neptune is 165 Earth Years

If you lived on Neptune, you'd only celebrate your first birthday after 165 Earth years. Forget cake—better bring a time machine.

Fact #557: Comets Leave Behind Trails of Dust

Comets shed material as they pass near the Sun, leaving behind trails that create meteor showers when Earth passes through them. Cosmic glitter, anyone?

Fact #558: The Dark Side of the Moon Isn't Always Dark

The Moon rotates, so every part of it gets sunlight at some point. The "dark side" is just the side we never see from Earth. Sorry, Pink Floyd fans!

Fact #559: Some Galaxies Are Shaped Like Doughnuts

Ring galaxies, like Hoag's Object, are shaped like cosmic doughnuts. Sweet, sugary, and entirely made of stars.

Fact #560: Saturn's Rings Are Disappearing

Saturn's iconic rings are slowly being pulled into the planet by gravity. Don't worry, they'll still be around for millions of years, so there's time to enjoy them.

Fact #561: The Moon Was Formed by a Giant Collision

Scientists believe the Moon formed when a Mars-sized planet collided with Earth billions of years ago. The debris from the crash eventually came together to create our favorite nighttime companion.

Fact #562: Jupiter's Moon Europa Might Have Alien Life

Europa has an icy crust with a salty ocean underneath, and scientists think it could harbor alien life. Move over, Mars—Europa is the new hotspot for extraterrestrial speculation.

Fact #563: A Day on Pluto is Almost a Week Long

On Pluto, a single day lasts about 153 Earth hours. That's six and a half days—plenty of time to binge-watch every sci-fi show ever made.

Fact #564: There Are Planets Where It's Raining Rocks

On exoplanet COROT-7b, the surface is so hot that rocks vaporize and rain back down as pebbles. Umbrellas just won't cut it there.

Fact #565: There's a Planet That Could Be a Water World

Exoplanet GJ 1214b is thought to be covered in water, with a thick steamy atmosphere. If true, it's basically an intergalactic hot tub.

Fact #566: Earth Once Had Two Moons

Some scientists theorize that Earth may have had a second, smaller moon billions of years ago, which eventually crashed into the larger one. Talk about sibling rivalry!

Fact #567: Venus Is Covered in Volcanoes

Venus has over 1,600 volcanoes, more than any other planet in the solar system. It's the solar system's unofficial fire emoji.

Fact #568: The Milky Way is Warping

Our galaxy isn't flat—it's warped like a cosmic tortilla chip. The gravitational pull of nearby galaxies could be causing this funky shape.

Fact #569: You Could Fit Over a Million Earths in the Sun

The Sun is so massive that over a million Earths could fit inside it. That's a lot of room for all the things Earthlings would probably overpack.

Fact #570: There's a Black Hole That's 12 Billion Times the Size of the Sun

The black hole TON 618 is so massive it makes even the Sun look tiny. Good luck wrapping your head around that cosmic heavyweight.

Fact #571: Mars Has Seasons, Just Like Earth

Mars experiences seasons because its axis is tilted, just like Earth's. The main difference? Its winters last twice as long. Pack extra sweaters!

Fact #572: You'd Weigh Less on the Moon

Gravity on the Moon is only about 1/6th of Earth's gravity. So, if you weigh 180 pounds on Earth, you'd weigh just 30 pounds on the Moon. Cosmic weight loss goals, anyone?

Fact #573: The Closest Star System Has Three Suns

Alpha Centauri, the closest star system to Earth, has three stars. Any planets there would have triple sunrises and sunsets—a cosmic overachiever.

Fact #574: You Can See 2.5 Million Galaxies with a Telescope

With a strong enough telescope, you could spot up to 2.5 million galaxies. That's a lot of star-studded neighbors to invite to the cosmic party.

Fact #575: Space is Filled with Invisible Dark Matter

Dark matter makes up about 85% of the universe, but it doesn't emit or reflect light. Scientists know it's there, but what it is? That's still the ultimate cosmic mystery.

Fact #576: Some Stars "Pulse" Like Beating Hearts

Variable stars expand and contract, causing them to pulsate. It's like the universe's way of reminding us that it's alive and kicking.

Fact #577: The Largest Known Star Could Swallow Our Solar System

The star UY Scuti is so massive that it could fit the Sun and every planet in our solar system inside it. Bet it's the life of the galactic party.

Fact #578: Earth Has a "Second Moon" (Kind of)

A small asteroid called 3753 Cruithne is sometimes referred to as Earth's second moon because it follows a similar orbit. It's more like a clingy space rock, but we'll take it.

Fact #579: Mercury is Shrinking

Mercury has been slowly shrinking over billions of years due to its cooling core. It's like the planet is aging—and it's doing it gracefully.

Fact #580: Black Holes Can Evaporate

Thanks to something called Hawking radiation, black holes lose energy over time and can eventually evaporate. Even cosmic giants have an expiration date.

Fact #581: The Moon Has No Atmosphere

Without an atmosphere, the Moon has no weather or protection from meteoroids. That's why its surface is covered in craters—talk about rough terrain.

Fact #582: There's a Planet That Orbits Its Star Every 8 Hours

The exoplanet SWEEPS-10 orbits its star in just 8 hours, making it one of the fastest orbits ever discovered. Imagine celebrating your birthday three times a day!

Fact #583: You'd Weigh 28 Times More on a Neutron Star

If you stood on a neutron star, gravity would crush you under 28 times your body weight. Thankfully, you'd never actually make it there—because you'd be obliterated long before landing.

Fact #584: Saturn's Moon Titan Has Liquid Lakes

Titan, one of Saturn's moons, has lakes made of liquid methane and ethane. It's like a freezing version of Earth's water cycle but with way more flammable fluids.

Fact #585: Laika, the First Dog in Space

In 1957, Laika, a stray from Moscow, became the first living creature to orbit Earth aboard Sputnik 2. As it was a one-way mission, she tragically died a few hours in due to overheating.

Her sacrifice marked a key milestone in space exploration, paving the way for human spaceflight.

Fact #586: The Fastest Thing in the Universe is Light

Light travels at 186,282 miles per second, making it the speed limit of the universe. Sorry, spaceships—you'll never outrun it.

Fact #587: There's a Storm on Neptune That's Bigger Than Earth

Neptune has a massive storm called the Great Dark Spot, which is roughly the size of Earth. It's basically a hurricane that never quits.

Fact #588: A Spaceship Could Orbit a Black Hole Forever

If a spaceship entered the right orbit around a black hole, it could theoretically orbit forever. Just don't get too close to the event horizon!

Fact #589: Mars Has Dust Storms That Cover the Entire Planet

Mars experiences global dust storms that can last for weeks or even months. Imagine trying to clean up after that mess!

Fact #590: The Milky Way is 13.6 Billion Years Old

Our galaxy has been around for nearly as long as the universe itself. It's seen more cosmic drama than a season of reality TV.

Fact #591: The Sun Will Eventually Swallow Earth

In about 5 billion years, the Sun will expand into a red giant, engulfing Mercury, Venus, and possibly Earth. Talk about a fiery end to the solar system.

Fact #592: There's a Galaxy Shaped Like a Smiley Face

Galaxy cluster SDSS J1038+4849 appears to have a smiley face thanks to gravitational lensing. Even the universe has a sense of humor!

Fact #593: A Day on Neptune is Only 16 Hours

Despite being one of the farthest planets from the Sun, Neptune spins rapidly, making its days just 16 hours long. Quick days, long nights.

Fact #594: Some Meteorites Are Older Than Earth

Certain meteorites are over 4.5 billion years old, predating the formation of Earth itself. Talk about ancient space souvenirs.

Fact #595: Space Isn't Empty—It's Full of Invisible Energy

Even the vast emptiness of space is filled with dark energy, which makes up 68% of the universe. Space isn't so lonely after all.

Fact #596: Astronauts Can't Snore in Space

Without gravity, air doesn't get trapped in astronauts' throats, meaning snoring is virtually impossible in space. Finally, a snore-free sleep!

Fact #597: There Are More Planets Than Stars

Astronomers estimate that there are more planets than stars in the universe—so yes, planets are the true cosmic overachievers.

Fact #598: Jupiter's Moons Have Oceans Under Their Ice

Europa and Ganymede, two of Jupiter's moons, are believed to have massive subsurface oceans. Move over, Pacific—these might be the biggest oceans in the universe.

Fact #599: The Moon Was Once Part of Earth

The Moon formed from debris after a massive collision between Earth and another planet-sized object billions of years ago. Cosmic breakups are messy.

Fact #600: Space Is Expanding Faster Than the Speed of Light

While nothing within space can travel faster than light, the space between galaxies is expanding faster than light itself. This cosmic stretch is driven by dark energy, and it's why distant galaxies are zooming away faster than we can ever hope to catch up. The universe just loves breaking the rules!

POP QUIZ

Ready to test your knowledge about the strange, fun, and fascinating universe? Let's see if you're a space genius or if you'll need to book a one-way ticket to Astronomy 101!

Questions

1. **True or False:** Space smells like barbecue.
2. Which planet could float in a giant bathtub because of its low density?
3. **True or False:** A day on Venus lasts longer than a year on Venus.
4. What is the name of Mars's largest volcano, which is three times taller than Mount Everest?
5. **True or False:** Tears float as little bubbles in space because there's no gravity.
6. Which planet has rings that are slowly disappearing?
7. **True or False:** You'd weigh less on the Moon than on Earth.
8. **True or False:** The Sun contains 99.8% of the total mass in our solar system.
9. **True or False:** Black holes can release bursts of energy after devouring matter.
10. What planet has a storm called the Great Red Spot that has been raging for over 300 years?

Answers

1. **True**: Astronauts report that space smells like barbecue—smoky and metallic. Who knew spacewalks came with built-in BBQ vibes?
2. **Saturn**: Saturn's low density means it could float in a bathtub—if you had one big enough, that is!
3. **True**: Venus's day is longer than its year because it spins very slowly on its axis. Talk about a long workday!
4. **Olympus Mons**: Mars's Olympus Mons is the largest volcano in the solar system, making Earth's mountains look like tiny hills.
5. **True**: Without gravity, tears can't fall, so they form into floating bubbles that stick to your face. Space crying is truly unique!
6. **Saturn**: Saturn's iconic rings are slowly being pulled into the planet by gravity, but don't worry—they'll stick around for millions of years.
7. **True**: Gravity on the Moon is about 1/6th of Earth's gravity, so you'd weigh much less there. Time to bounce like an astronaut!
8. **True**: The Sun's immense size and density account for nearly all the mass in our solar system.
9. **True**: Black holes can release powerful bursts of energy, known as "burps," after consuming matter. Even cosmic giants need to let it out!
10. **Jupiter**: The Great Red Spot on Jupiter is a massive storm that's been churning for centuries. Earth would fit inside it with room to spare!

How did you do? Whether you nailed it or learned something new, the cosmos is full of surprises—and so are pop quizzes!

CHAPTER 7

Haunted Histories

The world is full of stories that send shivers down your spine—unexplained phenomena, ghostly encounters, and eerie legends that make you want to sleep with the lights on. Some of these tales are spooky, some are bizarre, and some are just downright chilling. So, grab a flashlight and get ready for a journey into the creepy corners of the world.

Fact #601: The "Most Haunted House in America" is in San Diego

The Whaley House in San Diego is said to be haunted by its former residents, including Thomas Whaley himself. Visitors report hearing footsteps, seeing shadowy figures, and even smelling phantom cigars. Ghostly house party, anyone?

Fact #602: The Ghostly Underground of Edinburgh

Beneath the streets of Edinburgh lies a network of abandoned vaults, said to be haunted by spirits of those who once lived there in squalor. Visitors report hearing whispers, footsteps, and even spotting shadowy figures in the eerie tunnels.

Fact #603: The Bell Witch Legend Spooked Andrew Jackson

The Bell Witch haunting in Tennessee terrified an entire family in the 1800s, and even President Andrew Jackson reportedly had an encounter with the spirit. He's said to have called her a "terrible witch" before fleeing.

Fact #604: A Ghost Ship Roams the High Seas

The Flying Dutchman is a legendary ghost ship said to sail the oceans, unable to make port. Sailors claim to see its eerie glow during storms—because nothing screams "comforting" like a haunted ship in bad weather.

Fact #605: Scotland's Greyfriars Kirkyard is Crawling with Ghosts

Greyfriars Kirkyard in Edinburgh is famous for its ghostly residents, including the vengeful spirit of George Mackenzie, also known as the "Mackenzie Poltergeist." Visitors report being scratched or shoved—proof that not all ghosts are friendly.

Fact #606: The Queen Mary Hotel is a Floating Haunt

The Queen Mary, a luxury ocean liner turned hotel, is said to be one of the most haunted places in the world. Guests have reported hearing ghostly children playing and seeing a phantom woman in white wandering the halls.

Fact #607: The Ghostly Hitchhiker Urban Legend is Universal

From the U.S. to Japan, ghostly hitchhiker stories abound. A driver picks up a stranger who vanishes from the car without a trace—spooky, and also a solid excuse to never offer rides to strangers.

Fact #608: The Tower of London is Haunted by a Headless Queen

Anne Boleyn, Henry VIII's ill-fated wife, is said to haunt the Tower of London, carrying her head under her arm. She's proof that even the afterlife doesn't offer a clean break from royal drama.

Fact #609: A Forest in Japan is Known as the Sea of Trees

Aokigahara Forest, near Mount Fuji, is infamous for its eerie silence and stories of paranormal activity. Even seasoned hikers claim to hear whispers or see shadowy figures. Maybe bring a buddy—and a map.

Fact #610: A Phantom Dog Protects a Welsh Village

In Wales, the legend of the Gwyllgi, or "Black Hound," tells of a ghostly dog with glowing red eyes that roams the countryside. It's part protector, part nightmare fuel.

Fact #611: The Ghost of Flight 401 Still Checks on Passengers

After Flight 401 crashcd in 1972, parts of the plane were salvaged and reused in other aircraft. Passengers and crew on those planes have reported ghostly visits from the deceased flight crew. Talk about customer service from beyond the grave.

Fact #612: The Phantom Clown Phenomenon is Worldwide

Reports of creepy clowns lurking in parks and playgrounds have popped up in various countries. Whether it's mass hysteria or the world's creepiest prank, clowns just keep getting scarier.

Fact #613: The Lady in White Ghost is Everywhere

From Europe to Asia to the Americas, legends of ghostly women in white mourning their lost loves are universal. These spirits are either tragic or terrifying, depending on how close they get.

Fact #614: The Amityville Horror Was Based on a Real House

The Amityville Horror story, about a family haunted by dark forces, was based on events reported at a house in New York. Whether it was ghosts or a clever hoax, the tale still gives people nightmares.

Fact #615: A Haunted Painting Makes People Sick

The painting *The Hands Resist Him*, also called the "eBay Haunted Painting," is rumored to curse anyone who owns it. People report strange illnesses and eerie phenomena around it. Good luck finding a buyer!

Fact #616: Mirrors Can Trap Spirits

Some cultures believe mirrors are portals to other realms, and that covering them during a loved one's death prevents spirits from getting trapped. Suddenly, your bathroom mirror feels a little more ominous.

Fact #617: A Ghost Monk Haunts England's Borley Rectory

Dubbed the most haunted house in England, Borley Rectory was home to phantom footsteps, ghostly monks, and even floating objects. Sounds like a great Airbnb listing—for ghosts.

Fact #618: The Chilling Legend of the Banshee

In Irish folklore, the Banshee is a wailing spirit who warns families of impending death. Hearing her scream is said to be a terrifying—and very final—omen.

Fact #619: The RMS Titanic Has Ghostly Visitors

Visitors to Titanic exhibits report seeing ghostly apparitions and hearing voices near artifacts. Even after a century, the tragedy still lingers in the paranormal realm.

Fact #620: A Haunted Doll Named Annabelle Inspired a Movie

The real Annabelle doll, locked in a glass case in a museum, is said to be cursed. Its alleged antics inspired *The Conjuring* movies. Keep your dolls in check, people.

Fact #621: The "Cursed" Phone Number That Nobody Keeps

A Bulgarian phone number was reportedly linked to multiple sudden deaths of its owners. The number, 0888-888-888, was retired permanently. Coincidence? Maybe, but no one's willing to test it.

Fact #622: The Dyatlov Pass Mystery Still Baffles Experts

In 1959, nine hikers in Russia mysteriously died in the Ural Mountains under bizarre circumstances, including strange injuries and no signs of struggle. Theories range from avalanches to UFOs. The truth? Still unknown.

Fact #623: Haunted Castles Are Everywhere in Europe

From Edinburgh Castle in Scotland to Château de Brissac in France, Europe is teeming with haunted castles. Visitors report ghostly apparitions, chilling whispers, and even phantom knights.

Fact #624: The Legend of the Green Lady

Many haunted locations, from castles to forests, have stories of a "Green Lady" ghost. She's said to be a woman betrayed or heartbroken, forever wandering in her green attire. Fashionable yet terrifying.

Fact #625: Crybaby Bridges Are Urban Legends Across the U.S.

Crybaby Bridge tales involve hearing ghostly cries of babies or mothers late at night. From Ohio to Texas, these bridges are spooky staples in urban folklore.

Fact #626: Shadow People Are a Universal Phenomenon

Reports of shadowy figures seen in the corner of your eye are common worldwide. Whether they're ghosts or tricks of the mind, they're enough to make anyone sleep with the lights on.

Fact #627: The "Skeletons in the Walls" Tale Turned Out True

In 2017, workers renovating an old house in Wales discovered actual human skeletons in the walls. Legends of ghostly wailing were suddenly a lot more believable.

Fact #628: Ghost Trains Haunt Tracks Around the World

From Canada to Japan, tales of ghost trains abound. These spectral locomotives are said to appear late at night, often with no passengers or crew. All aboard for the creepiest ride ever!

Fact #629: The Ghost of Abraham Lincoln Haunts the White House

Multiple U.S. presidents and guests, including Winston Churchill, have reported seeing the ghost of Abraham Lincoln in the White House. Honest Abe seems to still be watching over the nation.

Fact #630: A Ghostly Nun Haunts Borley Rectory

Borley Rectory, once called the "most haunted house in England," is famous for its ghostly nun. Legend says she was bricked up alive after a forbidden romance. Spooky and tragic.

Fact #631: Phantom Footsteps in Old Hotels Are a Common Complaint

From the Stanley Hotel in Colorado to the Fairmont Banff Springs Hotel in Canada, guests frequently report hearing phantom footsteps in empty hallways. A ghostly concierge, perhaps?

Fact #632: The Mothman Legend Still Looms Large

In the 1960s, people in Point Pleasant, West Virginia, reported seeing a winged, red-eyed creature called the Mothman. It's said to appear before disasters, adding an eerie twist to its legend.

Fact #633: The Haunted Island of Poveglia

Located in Italy, Poveglia Island is considered one of the most haunted places on Earth, thanks to its history as a quarantine site and asylum. Even the locals won't go near it.

Fact #634: Ghost Lights in the Swamps

In various swamps worldwide, people report seeing glowing orbs of light called "will-o'-the-wisps." Some say they're mischievous spirits trying to lead travelers astray.

Fact #635: The Haunted Cultural Center of the Philippines

The Cultural Center of the Philippines (CCP), commissioned by Imelda Marcos and completed in 1969, is rumored to be haunted. According to legend, several workers were accidentally buried in wet cement during its rushed construction. Stories of ghostly apparitions and strange noises have added to its mystique, making it a popular subject of local folklore.

Fact #636: The Haunting at Leap Castle

Leap Castle in Ireland is said to be home to an "Elemental," a shadowy, malevolent spirit. Visitors report feelings of dread and even the smell of rotting flesh. Sounds like a hard pass.

Fact #637: Dolls Are a Common Source of Haunting Stories

From Annabelle to Robert the Doll in Florida, creepy dolls are often linked to supernatural tales. Why? Because nothing says "nightmare fuel" like a doll staring back at you.

Fact #638: The Ghost of a Pirate Haunts Ocracoke Island

The ghost of Blackbeard the pirate is said to haunt the shores of Ocracoke Island, where he was killed in 1718. Locals report seeing his ghostly figure and hearing phantom cannon fire.

Fact #639: The Haunting of the Myrtles Plantation

The Myrtles Plantation in Louisiana is rumored to be haunted by at least 12 ghosts, including Chloe, a former enslaved woman whose spirit is said to appear in photographs.

Fact #640: Ghostly Soldiers Haunt Gettysburg Battlefield

Visitors to Gettysburg often report seeing ghostly soldiers and hearing cannon fire and battle cries. Even after more than 150 years, it seems the spirits haven't left the field.

Fact #641: The Curse of the Crying Boy Painting

A series of house fires in England were reportedly linked to a painting called *The Crying Boy*. Despite the fires, the painting itself always remained untouched. Creepy or coincidence?

Fact #642: The Lemp Mansion is a Haunted Hotspot

In St. Louis, Missouri, the Lemp Mansion is said to be haunted by members of the tragic Lemp family, who faced financial ruin and suicides. Visitors report cold spots and ghostly voices.

Fact #643: The Devil's Footprints Mystery

In 1855, strange hoof-like prints appeared in the snow across England, stretching for miles. No one could explain them, sparking theories from the Devil to aliens.

Fact #644: Phantom Children Haunt the Villisca Axe Murder House

The Villisca Axe Murder House in Iowa, the site of a gruesome 1912 murder, is rumored to be haunted by the spirits of the children who died there. Visitors report hearing giggles and whispers.

Fact #645: The Hotel del Coronado's Ghostly Guest

The Hotel del Coronado in California is said to be haunted by the ghost of Kate Morgan, a guest who died under mysterious circumstances. She's been spotted wandering the hotel halls.

Fact #646: The Phantom Crier of Chillingham Castle

Chillingham Castle in England is home to a ghostly figure known as the "White Pantry Ghost," who is said to cry out for water. Maybe she just needs a better castle service.

Fact #647: The Ghost Fleet of Truk Lagoon

In Micronesia's Truk Lagoon, over 50 sunken WWII ships create eerie underwater scenes. Divers report spooky shadows and strange noises—ghosts, or just fish playing tricks?

Fact #648: A Poltergeist Terrorized a German Family in the 1960s

The Rosenheim Poltergeist case in Germany involved unexplained phone calls, flickering lights, and flying objects in a law office. Scientists and police investigated but couldn't explain the events, leaving it as one of the most documented hauntings ever.

Fact #649: The Vanishing City of Biringan

In the Philippines, the mythical Biringan City is said to be a hidden, otherworldly metropolis that only appears to chosen individuals. Stories describe it as a place of dazzling beauty, but those who enter might never return. Locals claim it's home to mystical beings called engkanto, making it a captivating blend of mystery and folklore.

Fact #650: The Haunting of Eastern State Penitentiary

This abandoned prison in Pennsylvania is said to be haunted by the spirits of former inmates. Visitors report ghostly whispers, shadowy figures, and the sound of cell doors slamming shut.

Fact #651: The Ghost of the Brown Lady of Raynham Hall

The Brown Lady, famously photographed in 1936, is said to haunt Raynham Hall in England. The spectral figure, supposedly Lady Dorothy Walpole, appears as a glowing silhouette in a brown dress.

Fact #652: Haunted Toys Are a Thing

A haunted Elmo doll in Florida reportedly began chanting "Kill James" without batteries installed. Whether it was a prank or paranormal, it's enough to give anyone nightmares.

Fact #653: The Ghost of Banff Springs Hotel

In Canada, Banff Springs Hotel is rumored to be haunted by several spirits, including a bride who tragically fell down the stairs. She's said to appear in her wedding dress, still waiting for her big day.

Fact #654: The Specter of Room 217 at the Stanley Hotel

The Stanley Hotel in Colorado, the inspiration for The Shining, has its fair share of hauntings. Room 217 is said to be haunted by a maid who was injured in a gas explosion. Guests report items being moved, lights flickering, and her ghostly presence.

Fact #655: The Bell Witch Cave

The Bell Witch haunting in Tennessee became so famous that the cave on the family's property is now a tourist attraction. Visitors report eerie whispers and even ghostly touches in the darkness.

Fact #656: Highgate Cemetery's Phantom Residents

Highgate Cemetery in London is not only a historic burial ground but also a hotspot for paranormal activity. Ghostly monks and eerie figures have been reported wandering the grounds.

Fact #657: The Haunted Ipatiev House

Before it was demolished, locals claimed the Ipatiev House in Yekaterinburg, particularly the basement where the Romanov family was executed, was haunted by their restless spirits. Witnesses reported hearing cries, whispers, and unexplained footsteps, adding to the house's eerie reputation.

Fact #658: The Phantom of the Opera House in Paris

The Palais Garnier opera house inspired the *Phantom of the Opera*, thanks to legends of ghostly sightings and strange occurrences in its underground lake. Spooky music, anyone?

Fact #659: The Ghosts of the Alamo

The Alamo, the site of a historic battle in Texas, is said to be haunted by the spirits of soldiers who died there. Visitors report hearing battle cries and seeing ghostly figures pacing the grounds.

Fact #660: The Haunted Road in Malaysia

Karak Highway is infamous for ghost sightings, including a mysterious woman in white and a phantom baby crying. Drivers are advised to avoid stopping—just in case.

Fact #661: The "Lady in Black" of Hollywood Forever Cemetery

In Los Angeles, a mysterious woman in black regularly visits Rudolph Valentino's grave, leaving flowers. Whether she's a fan or something more ghostly, no one knows for sure.

Fact #662: The Spectral Miners of the Coal Mines

Abandoned coal mines in the U.S. are said to be haunted by the ghosts of miners who perished in tragic accidents. Visitors report hearing the clang of tools and ghostly moans deep underground.

Fact #663: Ghost Cats Are a Thing

Ghostly cats have been reported in places like the British Museum and Edinburgh Castle. These spectral felines are said to roam their favorite spots, even in the afterlife.

Fact #664: The Headless Horseman of Sleepy Hollow

Washington Irving's famous tale was inspired by old legends of spectral horsemen, particularly in the Hudson Valley area. The headless rider still spooks visitors every Halloween.

Fact #665: A Ghostly Drummer Haunts Edinburgh Castle

The phantom drummer of Edinburgh Castle is said to appear before major events. Visitors report hearing drumming echoing through the stone halls, though no one is ever there.

Fact #666: The Devil's Number and Spooky Sightings

The number 666 is often associated with evil and paranormal activity. Coincidences involving this number are enough to make anyone's hair stand on end—just ask your receipt total.

Fact #667: The Ghost of a Monk Haunts Eynsford Castle

Visitors to this English castle have reported sightings of a ghostly monk who appears at dusk. Whether he's guarding secrets or just lost, he's a regular at this historic haunt.

Fact #668: Haunted Hospitals Are Real

Many abandoned hospitals, like Waverly Hills Sanatorium in Kentucky, are rumored to be hotspots for paranormal activity. Ghostly patients and eerie screams are frequently reported.

Fact #669: The "Ghost Boy" of 3 Men and a Baby

A scene in the movie *3 Men and a Baby* allegedly shows a ghostly boy standing in the background. It turned out to be a cardboard cutout, but the legend persists as one of Hollywood's creepiest bloopers.

Fact #670: The Curse of Lake Lanier

Lake Lanier in Georgia has a reputation for being haunted due to a history of accidents and drownings. Locals claim to see ghostly figures walking on the water at night.

Fact #671: The Lady in White of Union Cemetery

Union Cemetery in Connecticut is home to a famous Lady in White ghost, who is often seen gliding between tombstones. Drivers claim she even appears in the middle of the road at night.

Fact #672: The Haunting of the USS Hornet

The USS Hornet, a retired aircraft carrier, is said to be haunted by the spirits of sailors who served aboard. Visitors report seeing ghostly apparitions and hearing phantom footsteps.

Fact #673: The Ghostly Choir of Winchester Cathedral

Winchester Cathedral in England is said to echo with ghostly hymns sung by unseen choirs. The acoustics are heavenly— and haunting.

Fact #674: The Haunted Plantation in Georgia

The Sorrel-Weed House in Savannah is rumored to be one of the most haunted homes in the U.S. Visitors report seeing ghostly figures and hearing eerie laughter.

Fact #675: Ghostly Orbs in Photographs

While many believe ghostly orbs in photos are spirits, scientists often attribute them to dust or light reflections. Either way, they add an extra layer of mystery.

Fact #676: The Cursed Movie Set of The Exorcist

During the filming of *The Exorcist*, mysterious fires and injuries plagued the set. The crew even had a priest bless the production. Coincidence? Maybe. Creepy? Definitely.

Fact #677: Phantom Lights on Brown Mountain

The Brown Mountain Lights in North Carolina are unexplained glowing orbs that appear in the mountains. UFOs? Ghosts? Swamp gas? Nobody knows, but they're definitely spooky.

Fact #678: The Cursed Village of Kuldhara

Kuldhara, a village in India, was abandoned overnight 200 years ago. Legend says a curse dooms anyone who tries to live there, and it remains eerily empty to this day.

Fact #679: The Curse of the Basano Vase

An Italian vase from the 15th century is said to have brought death to every person who owned it. Eventually, it was buried to stop the curse—just in case.

Fact #680: The Ghost of Sarah's Attic

In a historic home in New Orleans, Sarah, a servant from the 1800s, is said to haunt the attic. Visitors report hearing faint whispers and the sound of furniture moving, despite no one being up there.

Fact #681: The Bridgewater Triangle is a Hotbed of Paranormal Activity

This area in Massachusetts is infamous for UFO sightings, mysterious creatures, and even cursed Native American burial grounds. It's like the Bermuda Triangle's spooky cousin.

Fact #682: The Legend of the White Lady of Uniondale

In South Africa, drivers on the Uniondale Road report picking up a woman in white who vanishes mid-ride. She's said to be the ghost of a woman who died in a car accident. Creepy rideshare, anyone?

Fact #683: The Ghost of the Eiffel Tower

Legend has it that a maintenance worker who tragically fell from the Eiffel Tower in the early 1900s haunts the landmark. Visitors sometimes report feeling an icy chill at night—just when the lights are their brightest.

Fact #684: The Screaming Tunnel of Niagara Falls

A tunnel near Niagara Falls is said to be haunted by the ghost of a young girl who died in a fire. Local legend claims you can hear her screams if you light a match inside the tunnel.

Fact #685: The Ghost of the Paris Catacombs

The Paris Catacombs, holding the remains of over six million people, are home to ghostly sightings and eerie sounds. Adventurers who wander off the guided path often report strange phenomena—or simply get lost forever.

Fact #686: The Phantom Drummer Boy of the Civil War

During the U.S. Civil War, soldiers reported hearing a phantom drummer boy playing during the night. His ghostly rhythm was said to predict an impending battle.

Fact #687: The Specter of the Haunted Piano

In an old house in New Orleans, a ghostly piano plays itself in the dead of night. No one ever sees who's playing, but the eerie melodies keep guests awake.

Fact #688: The Ghost of the Opera House in Australia

The Princess Theatre in Melbourne is haunted by the ghost of Frederick Federici, an opera singer who died on stage. His spectral figure is said to appear during rehearsals, keeping an eye on the performers.

Fact #689: The Legend of La Llorona

In Latin American folklore, La Llorona, or "The Weeping Woman," is said to haunt rivers, crying for her lost children. Hearing her wail is considered a bad omen—and reason to run.

Fact #690: The Haunted Lighthouse of St. Augustine

The St. Augustine Lighthouse in Florida is famous for ghostly activity, including the spirits of two young girls who tragically died there. Visitors report hearing giggles and seeing shadows darting up the stairs.

Fact #691: Phantom Lights on the Alaska Highway

Travelers on the Alaska Highway have reported mysterious lights that follow cars, only to disappear when approached. UFOs? Ghosts? Whatever it is, it's unnerving.

Fact #692: The Ghost Nun of Larnach Castle

Larnach Castle in New Zealand is said to be haunted by a ghostly nun who roams the halls. She's often seen near the grand staircase, silently watching visitors.

Fact #693: The Haunted Mines of Cornwall

The abandoned tin mines of Cornwall, England, are said to be haunted by the spirits of miners who died in accidents. Ghostly pickaxes and the sound of dripping water add to the eerie atmosphere.

Fact #694: The Curse of the Crying Cemetery

In a small cemetery in Mexico, visitors report hearing ghostly cries at night. Some believe it's the restless spirits of children, while others say it's La Llorona wandering the graves.

Fact #695: The Phantom Train of St. Louis

The St. Louis Light, a mysterious glowing orb along old train tracks, is believed to be the ghost of a railroad worker searching for his missing head. Spooky, yet oddly determined.

Fact #696: Haunted Mirrors Reflect More Than Just You

In old folklore, mirrors are said to trap the souls of the deceased. Some haunted hotels even cover mirrors to prevent ghostly apparitions from appearing in the glass.

Fact #697: The Ghost Ship of the Arctic

The HMS Terror and HMS Erebus, ships from a doomed Arctic expedition, have inspired tales of ghostly sightings. Modern explorers report eerie sounds and unexplained lights near their wreckage sites.

Fact #698: The Mystery of the Man in the Hat

Many people who experience sleep paralysis report seeing a shadowy figure wearing a hat. Whether he's a ghost, a demon, or a figment of the mind, his chilling presence is unforgettable.

Fact #699: The Bell Tower Ghost of Venice

In Venice, locals claim to see a ghostly figure ringing the bells of an abandoned tower. Whether it's a spirit or a clever prank, no one has been brave enough to climb up and find out.

Fact #700: The Ghost Who Texts Back

A bizarre modern ghost story involves people receiving text messages from deceased loved ones. While skeptics blame glitches, believers think it's the afterlife's way of staying connected.

POP QUIZ

Think you've got the courage to face these ghostly questions? Let's see how well you remember the chills and thrills from Chapter 7!

Questions

1. **True or False:** The Whaley House in San Diego is often called the most haunted house in America.

2. What is the name of the ghost ship said to roam the seas, glowing during storms?

3. What is the name of the real haunted doll that inspired The Conjuring movies?

4. Which haunted ship in California is said to have ghostly children playing in its halls?

5. **True or False:** The Tower of London is haunted by Anne Boleyn, who is sometimes seen carrying her own head.

6. What eerie forest in Japan is famous for its silence and tales of paranormal activity?

7. **True or False:** Shadow people are only reported in a few specific countries.

8. What famous cursed painting reportedly caused mysterious fires while remaining untouched?

9. **True or False:** The ghost of Abraham Lincoln has been spotted in the White House by multiple people.

10. Which Latin American ghost is known for crying near rivers while searching for her lost children?

Answers

1. **True:** San Diego's Whaley House is haunted by phantom footsteps, shadowy figures, and ghostly cigar smells.
2. **Flying Dutchman:** This legendary ghost ship is said to be cursed, doomed to roam the seas forever.
3. **Annabelle:** The cursed doll is locked in a glass case in a museum to prevent its alleged sinister antics.
4. **Queen Mary:** This ocean liner turned hotel is known for ghostly encounters, like playful children and a woman in white.
5. **True:** Anne Boleyn's ghost haunts the Tower of London, carrying her severed head.
6. **Aokigahara Forest:** Near Mount Fuji, this eerie forest is infamous for ghostly whispers and shadows.
7. **False:** Shadow people are reported worldwide, whether ghosts or strange illusions, they remain unsettling.
8. **Crying Boy Painting:** This artwork allegedly survived fires while homes burned around it—an eerie mystery.
9. **True:** Abraham Lincoln's ghost has been seen in the White House by many, including Winston Churchill.
10. **La Llorona:** This ghost from Latin American folklore haunts rivers, terrifying those who hear her cries.

How did you do? Whether you aced it or learned something new, these ghostly legends prove that the world of the paranormal is full of spooky surprises!

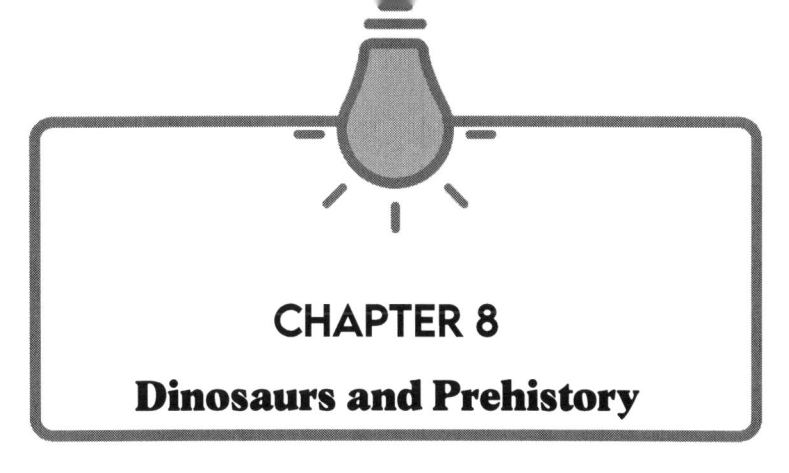

CHAPTER 8

Dinosaurs and Prehistory

Dinosaurs ruled the Earth long before humans even existed, and their world was as wild as it was gigantic. From T. Rex's tiny arms to the mystery of fossilized poop, dinosaurs are a constant source of fascination and fun. Get ready to dive into the prehistoric era where the facts are huge, and the delivery is anything but extinct!

Fact #701: The T. Rex Had Arms That Looked Like a Joke

T. Rex was one of the most fearsome predators of all time, but its tiny arms seemed hilariously out of proportion. Scientists believe they were strong enough to help hold prey or assist in getting back up—but they're still the punchline of countless dino jokes.

Fact #702: Some Dinosaurs Were the Size of Chickens

Not all dinosaurs were towering giants. Compsognathus, a speedy little carnivore, was roughly the size of a chicken. It's safe to say this dino wouldn't have made a scary Jurassic Park star.

Fact #703: Triceratops Was Basically a Walking Tank

With its massive horns and frill, Triceratops was well-equipped to fend off predators like T. Rex. Its frill may have also been used for showing off, making it the prehistoric version of a fancy hat.

Fact #704: The Stegosaurus Had a Brain the Size of a Lime

Despite its size, the Stegosaurus had a brain that weighed less than three ounces. This might explain why it didn't seem to mind having spikes on its tail instead of useful defense skills.

Fact #705: Some Dinosaurs Had Feathers

Yes, you read that right—many dinosaurs, including Velociraptors, had feathers. They weren't exactly ready for flight, but imagine a giant, scaly, feathered chicken running at you. Terrifying.

Fact #706: The Brachiosaurus Was Taller Than a Four-Story Building

Brachiosaurus could stretch its long neck to munch on treetops that were as tall as four-story buildings. It was like a prehistoric giraffe—if giraffes also weighed over 50 tons.

Fact #707: Ankylosaurus Was the Ultimate Armored Vehicle

Ankylosaurus was covered in thick, bony armor and wielded a club-like tail that could shatter bones. Think of it as the medieval knight of the dinosaur world.

Fact #708: The T. Rex Had Teeth the Size of Bananas

T. Rex's teeth were massive, sharp, and could crush bone. Each one was about the size of a banana, making it a true "chomping champion" of the prehistoric world.

Fact #709: Fossilized Dinosaur Poop is a Thing

Known as coprolites, fossilized dinosaur poop provides scientists with clues about what dinosaurs ate. Yes, paleontologists actually study ancient dino droppings for science—and it's oddly fascinating.

Fact #710: Dinosaurs Had Hollow Bones

Like modern birds, many dinosaurs had hollow bones that made them lighter and more agile. It's one of the reasons scientists believe birds are the closest living relatives of dinosaurs.

Fact #711: T. Rex Couldn't Run Fast

Despite being a terrifying predator, T. Rex wasn't built for speed. Scientists estimate it could run up to 12 mph—so, theoretically, you could outbike a T. Rex if you pedaled hard enough.

Fact #712: Dinosaurs Experienced Growth Rings in Their Bones

Like trees, some dinosaurs had growth rings in their bones, which scientists use to determine their age and growth patterns. These rings reveal fascinating details about their life spans and environmental conditions.

Fact #713: Spinosaurus Was the First Swimming Dinosaur

Spinosaurus is believed to have been semi-aquatic, hunting fish in rivers and swamps. With its massive sail-like spine, it probably looked like a prehistoric shark with legs.

Fact #714: Diplodocus Had a Whip-Like Tail

Diplodocus, one of the longest dinosaurs, could whip its tail at speeds of up to 30 mph. Imagine a tail so strong it could create a mini sonic boom. That's a tail worth staying away from.

Fact #715: Dinosaur Fossils Were Mistaken for Dragon Bones

Before paleontology became a thing, people often thought dinosaur fossils were the remains of dragons. It's easy to see how a T. Rex skull could spark some fiery legends.

Fact #716: The Velociraptor Wasn't That Big

Thanks to Hollywood, people imagine Velociraptors as giant predators, but in reality, they were about the size of a turkey. Feathers and all. Not exactly the scariest dino at Thanksgiving.

Fact #717: The Earth Looked Very Different in Dinosaur Times

During the age of dinosaurs, all the continents were connected in a giant supercontinent called Pangaea. It was the ultimate prehistoric party before they drifted apart.

Fact #718: The Pachycephalosaurus Used Its Skull Like a Battering Ram

This dinosaur had a thick, domed skull it used for head-butting rivals. Basically, Pachycephalosaurus invented extreme sports millions of years ago.

Fact #719: Sauropods Had Tiny Heads for Their Massive Bodies

Sauropods like Apatosaurus had long necks and massive bodies but hilariously small heads. Imagine a dinosaur with a head smaller than a horse's—it's like evolution played a prank.

Fact #720: Dinosaurs Didn't All Go Extinct

Birds are considered direct descendants of dinosaurs, which means your pet parrot is basically a tiny T. Rex in disguise. Just don't tell it—it might go full Jurassic.

Fact #721: The Microraptor Was the Dinosaur That Could Glide

Microraptor was a small, feathered dinosaur with four wings that allowed it to glide from tree to tree. It's the prehistoric ancestor of hang gliders, but way cooler.

Fact #722: Carnivorous Dinosaurs Had Hollow Teeth

The sharp teeth of carnivorous dinosaurs like T. Rex were hollow, which made them lighter and more efficient for chomping prey. Less weight, more bite—nature's perfect design.

Fact #723: Some Dinosaurs Lived in the Arctic

Fossil discoveries in Alaska suggest that some dinosaurs adapted to cold, dark winters. Imagine a snow-dwelling dino— no mittens required!

Fact #724: Archaeopteryx Was a Feathered Dinosaur That Could Fly

Archaeopteryx, often called the first bird, had feathers and wings but still retained dinosaur-like features. It's proof that evolution doesn't rush its projects.

Fact #725: Dinosaurs Came in All Shapes and Sizes

From the tiny mouse-sized Lesothosaurus to the gigantic Argentinosaurus (the size of three school buses), dinosaurs were the ultimate size diversity squad.

Fact #726: T. Rex Could Crush Bones Like Potato Chips

With a bite force of 12,000 pounds, T. Rex could crunch through bones like they were chips at a party. It's safe to say that "sharing snacks" wasn't its strong suit.

Fact #727: Parasaurolophus Had a Built-In Horn

This dinosaur had a long, hollow crest on its head that worked like a trumpet. Scientists think it used this for communication, scaring predators, or even dino karaoke nights.

Fact #728: Dinosaur Eggs Were Surprisingly Small

Even the largest dinosaurs laid eggs that were no bigger than a football. Imagine the size of a dino omelet, though—now that's breakfast goals.

Fact #729: The Fastest Dinosaur Was Ornithomimus

Ornithomimus, also known as the "bird mimic," could run up to 40 miles per hour. Think of it as the Usain Bolt of the dinosaur world, but with feathers.

Fact #730: Some Dinosaurs Had Beaks

Just like birds, certain dinosaurs like Protoceratops had beaks instead of teeth. They were great for munching plants, but not so good for dino dentist visits.

Fact #731: Fossilized Skin Imprints Show Dino Scales

Some dinosaur fossils have preserved skin imprints, revealing that many dinosaurs had scaly skin similar to reptiles. No moisturizers needed in the prehistoric world.

Fact #732: Argentinosaurus Was the Biggest Dino of Them All

Argentinosaurus was the largest dinosaur ever discovered, measuring up to 100 feet long. This herbivore was basically a walking skyscraper with a leafy appetite.

Fact #733: Dinosaur Names Are Like a Prehistoric Vocabulary Test

Most dinosaur names come from Greek or Latin, like *Triceratops* (three-horned face) or *Velociraptor* (swift thief). If you can pronounce them all, you deserve a medal.

Fact #734: T. Rex Had Great Vision

Despite the myth that T. Rex couldn't see you if you stood still, its binocular vision was excellent. Hiding from one would have been about as effective as playing peek-a-boo with your dog.

Fact #735: Pterosaurs Could Glide for Miles

While not technically dinosaurs, pterosaurs ruled the skies of the prehistoric world. Some species, like Quetzalcoatlus, had wingspans as wide as a small airplane.

Fact #736: Dinosaur Herds Were Like Prehistoric Road Trips

Evidence suggests that some dinosaurs, like hadrosaurs, traveled in herds for protection and socializing. Picture a dino caravan stomping across ancient landscapes.

Fact #737: The First Dinosaur Fossil Was Mistaken for a Giant Lizard

When the first dinosaur fossil was discovered in 1676, scientists thought it belonged to a giant lizard. It wasn't until the 1800s that the term "dinosaur" was coined, meaning "terrible lizard."

Fact #738: Dinosaurs Got Replaced by Mammals

When the dinosaurs went extinct 66 million years ago, mammals took over the planet. If not for that asteroid, Earth might still be a dinosaur's playground.

Fact #739: Diplodocus Could Use Its Tail as a Whip

Diplodocus had a long, flexible tail that it could snap like a whip. Some scientists believe it used this to scare off predators or even communicate with other dinos.

Fact #740: Some Dinosaurs Had Built-In Air Conditioning

Certain large dinosaurs had hollow bones and special air sacs that helped them stay cool in hot climates. Prehistoric engineering at its finest!

Fact #741: Spinosaurus Had a Crocodile-Like Snout

Spinosaurus had a long, narrow snout filled with sharp teeth, perfect for catching fish. Imagine a dino that could double as a fishing expert—talk about versatility.

Fact #742: Dinosaur Tracks Are Like Ancient Selfies

Fossilized dinosaur footprints give scientists clues about their movement, behavior, and even social habits. It's like prehistoric Instagram, but way cooler.

Fact #743: The Mass Extinction Was the Ultimate Game Over

About 66 million years ago, an asteroid wiped out the dinosaurs. The impact created massive fires, tsunamis, and a global winter. If you think bad weather ruins your day, imagine that.

Fact #744: Dinosaurs Had Great Social Lives

Evidence shows that many dinosaurs were social animals, living in groups and even helping care for their young. The original squad goals, prehistoric style.

Fact #745: Some Dinosaurs Could Swim

Dinosaurs like Spinosaurus and Baryonyx were excellent swimmers, diving into rivers and lakes for food. They were the Michael Phelps of the prehistoric age.

Fact #746: Dino Teeth Grew Back Quickly

Carnivorous dinosaurs like T. Rex could lose teeth in a fight or while eating, but they'd grow back in a matter of weeks. It's like having a never-ending supply of steak knives.

Fact #747: Scientists Are Still Discovering New Dinosaurs

More than 1,000 dinosaur species have been identified, but new ones are discovered every year. Dinosaurs may be extinct, but they're not done surprising us.

Fact #748: Prehistoric Crocodiles Were Just as Scary as Dinosaurs

During the age of dinosaurs, giant prehistoric crocodiles like Deinosuchus grew up to 40 feet long and could take down even the fiercest dinos. Talk about a prehistoric villain.

Fact #749: Fossilized Amber Preserved Prehistoric Insects

Amber, fossilized tree resin, has trapped ancient insects, plants, and even small animals. It's like nature's time capsule, and yes, it inspired *Jurassic Park*.

Fact #750: Dinosaurs Were Colorful

While many people imagine dinosaurs in dull colors, scientists believe they were likely as colorful as modern birds. Picture a rainbow T. Rex—it's not as far-fetched as you'd think.

Fact #751: T. Rex Had Built-In Night Vision

T. Rex had forward-facing eyes, giving it excellent depth perception and night vision. If it hunted you at night, hiding under your bed wouldn't save you.

Fact #752: Dinosaurs Could Lose Teeth Like Sharks

Carnivorous dinosaurs like T. Rex and Allosaurus continuously replaced their teeth. Lose a tooth in a fight? No problem—just grow a new one and keep chomping.

Fact #753: The Largest Dino Egg Ever Found Was the Size of a Basketball

Titanosaurs, some of the largest dinosaurs ever, laid eggs about the size of a basketball. That might not seem huge, but it's impressive for something that could grow up to 100 feet long!

Fact #754: Pterosaurs Had Fluffy Coats

Some pterosaurs, like Dimorphodon, had hair-like structures to help keep them warm. Imagine a flying dinosaur in a fuzzy jacket—it's as adorable as it is terrifying.

Fact #755: Stegosaurus Might Have Used Its Plates to Communicate

The plates on a Stegosaurus's back may have changed color to scare off predators or attract a mate. Think of them as prehistoric mood rings.

Fact #756: Carnivorous Dinosaurs Ate Bones

Dinos like T. Rex didn't just eat meat—they crunched up bones too. Scientists know this because they've found bone fragments in fossilized dinosaur poop. Talk about commitment to a meal.

Fact #757: Some Dinosaurs Had Tiny Arms—For a Reason

Carnotaurus, a predator with even tinier arms than T. Rex, probably used them for balance or mating displays. Their arms might have been small, but their attitude was big.

Fact #758: Herbivorous Dinosaurs Were Gigantic Poop Machines

Dinosaurs like Brachiosaurus had to eat hundreds of pounds of plants daily, producing an equally impressive amount of, well, fertilizer. Imagine the cleanup crew required for that mess.

Fact #759: Dinosaurs May Have Had Great Hearing

Some dinosaurs, like hadrosaurs, had crests that amplified sounds. They could communicate over long distances—like a prehistoric walkie-talkie system.

Fact #760: Troodon Had a Brain Bigger Than Most Dinos

Troodon was a small, bird-like dinosaur with a relatively large brain for its size. It wasn't writing poetry, but it may have been one of the smartest dinos on the block.

Fact #761: Dinosaurs Had Really Weird Teeth

Some herbivorous dinosaurs, like Nigersaurus, had over 500 teeth arranged like a lawnmower. Perfect for mowing down prehistoric vegetation!

Fact #762: T. Rex Had Built-In Air Conditioning

T. Rex had large air pockets in its skull that helped keep its head cool. Even the king of the dinosaurs needed to chill sometimes.

Fact #763: Iguanodon Was the First Dino Identified

Iguanodon was one of the first dinosaurs ever discovered, identified in 1825. Its fossilized thumb spike became a symbol of early paleontology—and a prehistoric weapon.

Fact #764: Some Dinosaurs Had Built-In "Safety Helmets"

Pachycephalosaurus had a thick, domed skull that was over 10 inches thick. Scientists believe they used it to headbutt each other during dino disputes—or just to show off.

Fact #765: Dino Fossils Are Older Than the Himalayas

Many dinosaur fossils predate the formation of the Himalayas, which are only about 50 million years old. Dinosaurs really put "old" in "old school."

Fact #766: Some Dinosaurs Were Gluttons for Rocks

Dinosaurs like Apatosaurus swallowed small stones called gastroliths to help grind up tough plants in their stomachs. Nature's blender was a handful of pebbles.

Fact #767: Dinosaurs Had Weird Tails

Some dinosaurs, like the Ankylosaurus, had clubbed tails that could shatter bones, while others, like Diplodocus, had whip-like tails that could make a sonic boom. Functional and fashionable!

Fact #768: T. Rex Was a Kid for a Long Time

T. Rex didn't reach full size until about 20 years old, spending much of its life as a lanky teenager. Imagine a gangly teenage T. Rex with braces. Terrifying and awkward.

Fact #769: Pterosaurs Were Not Dinosaurs

Although they lived during the same time, pterosaurs were flying reptiles, not dinosaurs. Think of them as the cool cousins who crashed the dino party.

Fact #770: Dinosaurs Lived on All Continents

Fossils have been found on every continent, including Antarctica. Even the coldest places on Earth were once warm enough for dinosaurs to roam.

Fact #771: There Were Armored Dinosaurs Galore

Besides Ankylosaurus, there were other tank-like dinos, such as Scelidosaurus, covered in bony plates and spikes. They were like prehistoric battle tanks with attitude.

Fact #772: T. Rex's Closest Relative is a Chicken

Scientists have found that modern chickens share DNA with T. Rex. So the next time you see a chicken, just remember—it's a distant, feathery descendant of the king of the dinosaurs.

Fact #773: Velociraptor Had a Killer Claw

Velociraptor had a large, curved claw on each foot that it used to slash prey. While it wasn't as big as its movie counterpart, it was still deadly and efficient.

Fact #774: Plant-Eating Dinos Traveled in Herds

Herbivorous dinosaurs like Maiasaura traveled in massive herds to protect their young from predators. Safety in numbers isn't just a human strategy.

Fact #775: Dinosaurs Had Colorful Feathers

Some dinosaurs, like Microraptor, had iridescent feathers that shimmered like a modern peacock. Evolution clearly had fun designing these prehistoric creatures.

Fact #776: Some Dinosaurs Could Glide

Small dinosaurs like Yi qi had bat-like wings that allowed them to glide between trees. They were nature's first experiment with flight.

Fact #777: The "Dino Wars" of the 1800s Were Intense

During the 19th century, two paleontologists, Cope and Marsh, competed fiercely to discover and name new dinosaurs. Their rivalry led to incredible discoveries—and some shady science.

Fact #778: Dinosaurs Were Really Into Recycling

Dinosaur nests were often reused by other animals after the dinos moved out. Even prehistoric creatures believed in reducing, reusing, and recycling.

Fact #779: Carnotaurus Had Tiny Horns

Carnotaurus, a fierce predator, had small horns above its eyes. While they probably weren't great for fighting, they definitely added a touch of flair to its look.

Fact #780: Dino Footprints Are Everywhere

Dinosaur footprints have been found on every continent, providing valuable insight into how they moved and lived. It's like the Earth was their personal scrapbook.

Fact #781: Diplodocus Could Wag Its Tail Like a Whip

Diplodocus had one of the longest tails of any dinosaur, and it could whip it so fast that it might have produced a sound similar to a gunshot. Prehistoric tail power, unlocked.

Fact #782: Dinosaurs Were Not the First Reptiles

Before dinosaurs ruled the Earth, prehistoric reptiles like Dimetrodon dominated the scene. These sail-backed creatures paved the way for the dinosaur takeover.

Fact #783: Triceratops Might Have Used Its Horns to Flirt

While its horns could fend off predators like T. Rex, Triceratops might have also used them to attract mates or show off dominance. Think of it as the ultimate prehistoric pickup line.

Fact #784: Sauropods Were Constantly Eating

With their enormous size, sauropods like Brachiosaurus had to eat up to 1,000 pounds of plants daily. It's a wonder they didn't eat the world out of its greenery.

Fact #785: Some Dinosaurs Had Long, Feathered Tails

Feathered dinosaurs like Caudipteryx had elaborate tails that were likely used for mating displays, much like modern peacocks. Fancy tail, prehistoric style.

Fact #786: The Hadrosaurus Was a Prehistoric Trumpet

Hadrosaurs, also known as duck-billed dinosaurs, had crests that acted like resonating chambers. They could honk, whistle, or trumpet to communicate across long distances. Dino jazz, anyone?

Fact #787: T. Rex Couldn't Chew

Despite its powerful jaws, T. Rex didn't chew its food. Instead, it tore massive chunks of meat and swallowed them whole. Table manners? Nonexistent.

Fact #788: Fossils Sometimes Contain Dino Skin

Rare fossils have preserved not just bones, but skin imprints, showing us exactly what dinosaurs looked like. It's like nature's version of a selfie.

Fact #789: Dinosaurs Got Hairy Toward the End

Some late-period dinosaurs were covered in proto-feathers, showing that evolution was already experimenting with flight before the asteroid hit.

Fact #790: Pterosaurs Could Walk Like Bats

When not flying, pterosaurs walked on all fours, much like bats. It's a reminder that prehistoric flying reptiles were as versatile as they were creepy.

Fact #791: Baby Dinosaurs Hatched with Teeth

Some dinosaur eggs hatched with babies ready to bite. Those tiny teeth helped them break out of their shells and start snacking on plants—or meat—right away.

Fact #792: Dino Nests Came in All Shapes and Sizes

From small, shallow scrapes to massive, carefully arranged circles, dinosaur nests varied as much as the dinos themselves. They were the original architects of the prehistoric world.

Fact #793: Velociraptors Had Sharp Toe Claws

Velociraptors are famous for their sickle-shaped toe claws, which were their primary weapons. It's not hard to see why they've been Hollywood's favorite predators.

Fact #794: Not All Dinosaurs Were Huge

While giants like Argentinosaurus stole the spotlight, many dinosaurs were smaller than humans, including some that were only a few feet tall. Dinosaurs came in all shapes and sizes!

Fact #795: T. Rex Was a Terrifying Teenager

During its teenage years, T. Rex went through a growth spurt, gaining up to 5 pounds a day. Imagine an awkward, hungry, and hormonal dino stomping around.

Fact #796: Armored Dinosaurs Had Built-In Defense Systems

Dinosaurs like Ankylosaurus had bony armor and club tails to fend off predators, making them prehistoric tanks. Good luck getting through that shell!

Fact #797: Dino Teeth Were Self-Sharpening

Some dinosaurs had teeth with ridges that naturally sharpened themselves as they ate. It's like having built-in dental tools.

Fact #798: The Largest Dino Footprint Was 5 Feet Wide

Discovered in Australia, the largest dinosaur footprint ever found measures over 5 feet wide. Imagine trying to outrun something that big—it's not happening.

Fact #799: Some Dinosaurs Had Quirky Head Shapes

Dinosaurs like Parasaurolophus had head crests that looked like they belonged in a sci-fi movie. Scientists think they were used for communication, but we like to think they were just showing off.

Fact #800: Dinosaurs Were the Original Survivors

Dinosaurs dominated the Earth for an incredible 160 million years, enduring volcanic eruptions, shifting continents, and dramatic climate changes. By comparison, humans have only been around for about 300,000 years—a mere blink in the timeline of life on Earth. If life were a movie, dinosaurs would be the epic trilogy, and humans would be the post-credits scene!

POP QUIZ

Let's see how much you remember about these prehistoric giants! Ready to stomp through this dino-themed quiz?

Questions

1. **True or False:** T. Rex had tiny arms that were completely useless.
2. What dinosaur had a tail that could whip so fast it might have created a sonic boom?
3. **True or False:** Some dinosaurs were the size of chickens.
4. What was the purpose of the Stegosaurus's back plates?
5. **True or False:** Velociraptors were as big as the ones in the movies.
6. What massive dinosaur was so big it could grow up to 100 feet in length, making it one of the largest ever discovered?
7. **True or False:** Dinosaurs lived only on warm continents and never in colder climates like Antarctica.
8. What dinosaur is known for having a domed skull that it likely used for headbutting?
9. **True or False:** Birds are considered living descendants of dinosaurs.
10. What was the purpose of the small stones (gastroliths) that dinosaurs like Apatosaurus swallowed?

Answers

1. **False**: T. Rex's arms were small, but they were incredibly strong and likely helped hold prey or push the dinosaur up when it fell. They weren't useless—just short!
2. **Diplodocus**: Diplodocus had a long, whip-like tail that it could snap at high speeds, potentially creating a sound similar to a gunshot.
3. **True**: Dinosaurs like Compsognathus were about the size of chickens. Not every dino was a towering giant!
4. **Communication and Defense**: Stegosaurus's plates may have been used to regulate body heat, attract mates, or intimidate predators—not for direct combat.
5. **False**: Real Velociraptors were about the size of a turkey and were covered in feathers. Hollywood made them much scarier.
6. **Argentinosaurus**: Argentinosaurus was one of the largest dinosaurs ever, reaching lengths of up to 100 feet. It's the true heavyweight champ of the dinosaur world.
7. **False**: Fossils have been found on all continents, including Antarctica. Some dinosaurs were adapted to colder climates.
8. **Pachycephalosaurus**: This thick-skulled dinosaur likely used its domed head for headbutting rivals or impressing potential mates.
9. **True**: Birds are considered modern-day dinosaurs, evolving directly from theropod dinosaurs like Velociraptors.
10. **Grinding Food**: Dinosaurs swallowed gastroliths to help grind up tough plants in their stomachs. These small stones acted like natural food processors.

How did you do? Whether you're a dino expert or learned something new, prehistoric trivia never goes extinct!

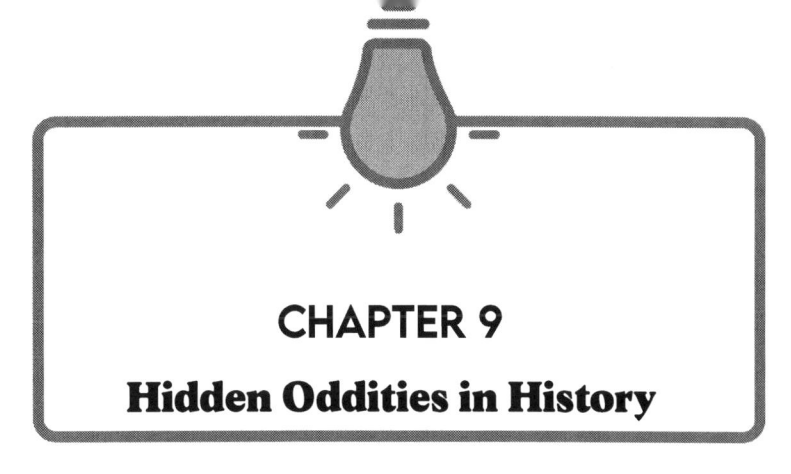

CHAPTER 9
Hidden Oddities in History

History isn't just about dusty books and old statues. It's packed with weird, funny, and downright bizarre moments that will make you question everything you thought you knew. From accidental discoveries to strange traditions, this chapter dives into the wacky side of the past.

Fact #801: Napoleon Was Attacked by a Horde of Bunnies

During a hunting trip, Napoleon Bonaparte was ambushed— not by enemies, but by rabbits! His staff had released hundreds of them for the hunt, but instead of fleeing, the bunnies charged at him. The emperor of France was defeated by fluffy insurgents.

Fact #802: Cleopatra Lived Closer to the Moon Landing Than to the Pyramids

The Great Pyramid of Giza was built around 2,500 BCE, while Cleopatra ruled in 51 BCE. That means she's closer in time to Neil Armstrong walking on the Moon than she is to the building of Egypt's iconic monuments.

Fact #803: Medieval Dentists Used Live Worms

Back in the Middle Ages, if you had a toothache, a common remedy was placing a live worm on the sore tooth to "eat" the pain. If that sounds gross, just remember—it was long before toothpaste was invented.

Fact #804: A Chicken Survived Without Its Head

In 1945, a chicken named Mike survived for 18 months after his head was cut off. His owner accidentally missed the brain stem, and Mike became a headless celebrity, touring the country. He even had his own tiny drip feeder for food.

Fact #805: Julius Caesar Was Once Kidnapped by Pirates

When pirates captured Julius Caesar, he was insulted by the ransom amount they demanded—so he demanded they raise it. After being freed, Caesar returned with a fleet, captured the pirates, and had them executed. Don't mess with Caesar.

Fact #806: A War Was Fought Over a Pig

In 1859, the U.S. and Britain nearly went to war over the death of a pig on the border between Canada and the U.S. Luckily, cooler heads prevailed, and the "Pig War" ended without a single human casualty.

Fact #807: A King Died From Laughing Too Much

King Martin of Aragon reportedly died in 1410 after laughing uncontrollably at a joke about a deer. His death might not be ideal, but at least he went out with a smile.

Fact #808: People Used to Have Ice Cream Funerals

In Victorian England, wealthy families sometimes hired "ice cream servers" to hand out treats during funerals. Mourning is a little easier when you're holding a cone of mint chocolate chip.

Fact #809: Einstein's Brain Was Stolen

After Albert Einstein died, a doctor removed his brain without permission. It was kept in jars and studied for decades, but the secret to his genius remains elusive. Moral of the story: write a will.

Fact #810: Cats Helped Win World War II

British ships in WWII kept cats on board to control rats and boost morale. One famous feline, Unsinkable Sam, survived three shipwrecks and became a naval legend. Proof that cats really do have nine lives.

Fact #811: George Washington Was a Mule Breeder

The first U.S. president was obsessed with mules, considering them essential for farming. He bred dozens of them at his estate, Mount Vernon, making him the ultimate "founding farmer."

Fact #812: Ancient Romans Had Glow-in-the-Dark Stones

The Romans discovered a type of stone that glowed in the dark when exposed to sunlight. They had no idea it was radioactive, but hey, they were ahead of their time in spooky decor.

Fact #813: Queen Elizabeth I Banned Mirrors

Later in life, Queen Elizabeth I banned mirrors in her court. She didn't want to see the signs of aging, proving that even royals deal with bad hair days.

Fact #814: A 17-Year-Old Invented the Snowmobile

In 1922, a Canadian teenager named Joseph-Armand Bombardier invented the first snowmobile using a Model T engine. What were you doing at 17? Probably not inventing winter transportation.

Fact #815: Ancient Egyptians Loved Board Games

The Egyptians played a game called Senet, which was a mix of strategy and luck. It was so popular that even pharaohs were buried with game boards to play in the afterlife. Prehistoric gamers, unite!

Fact #816: Medieval Doctors Dressed Like Birds

Those creepy plague doctor costumes with long beak-like masks were meant to protect physicians from "bad air." The beaks were stuffed with herbs, making them the first air fresheners—kind of.

Fact #817: New York Used to Be Called New Amsterdam

Before it became New York, the city was known as New Amsterdam, a Dutch colony. Imagine saying, "I 🖤 New Amsterdam" instead—doesn't quite have the same ring to it.

Fact #818: The Eiffel Tower Was Almost Demolished

The Eiffel Tower was originally a temporary structure for the 1889 World's Fair. After the event, it was supposed to be torn down, but Parisians loved it so much they kept it. Good call, Paris.

Fact #819: A Swiss Guard Once Danced on the Pope's Head

During the 16th century, a mischievous Swiss Guard reportedly danced on a sleeping pope's head as a prank. It's unclear if he kept his job, but it's safe to assume the pope wasn't amused.

Fact #820: Ketchup Was Originally Medicine

In the 1830s, ketchup was sold as a cure for indigestion. It wasn't until later that someone thought, "Hey, this would be great on fries!"

Fact #821: A City Accidentally Baked Its Banknotes

In 1920s Ireland, a fire broke out at a bakery where cash was being stored temporarily. The banknotes were baked into crispy shapes but were still accepted as legal tender. Talk about money laundering.

Fact #822: Napoleon Was Short? Not Really!

The myth that Napoleon was unusually short came from confusion over French and English measurement systems. At about 5'6", he was average height for his time. Turns out his ego wasn't compensating for much.

Fact #823: Russians Once Tried to Train Dogs to Deliver Bombs

During World War II, Soviet forces trained dogs to carry bombs under enemy tanks. Unfortunately, the dogs often ran back to their own tanks instead, causing chaos.

Fact #824: Toilet Paper Was Invented in China

The Chinese invented toilet paper in the 6th century, but it was considered a luxury item. Imagine living in a time when TP was reserved for royalty—talk about roughing it.

Fact #825: Benjamin Franklin Invented Swim Fins

Before becoming a Founding Father, Franklin invented swim fins at age 11. They were worn on his hands and helped him glide through the water faster. Young Ben was way ahead of his time.

Fact #826: A Pope Once Declared War on Cats

Pope Gregory IX declared that cats were associated with witchcraft in the 13th century, leading to widespread cat killings. Ironically, fewer cats led to more rats, worsening the plague. Oops.

Fact #827: The Shortest War in History Lasted 38 Minutes

In 1896, Britain and Zanzibar went to war, and Zanzibar surrendered after just 38 minutes. It's the world's shortest war—barely long enough for a coffee break.

Fact #828: Thomas Edison Was Afraid of the Dark

The inventor of the light bulb reportedly had a fear of the dark. Talk about turning your phobia into motivation.

Fact #829: The U.S. Once Had a Camel Corps

In the 1850s, the U.S. Army experimented with using camels for transportation in desert regions. It worked well, but the project was abandoned because, well, they were camels, and soldiers didn't love them.

Fact #830: People Used to Believe Tomatoes Were Poisonous

In the 18th century, Europeans called tomatoes "poison apples" because people who ate them often got sick. The real culprit? Lead plates reacting with the tomato's acidity. Thankfully, now they're just delicious.

Fact #831: A Shark Attack Changed History

In 1916, a series of shark attacks along the New Jersey coast terrified beachgoers and inspired *Jaws*. It also led to the first organized efforts to study sharks, so you can thank those attacks for modern shark science.

Fact #832: A Man Survived Two Atomic Bombs

Tsutomu Yamaguchi was in Hiroshima during the first atomic bomb attack and survived. He then traveled to Nagasaki—just in time for the second bombing. He lived to be 93, proving he was basically indestructible.

Fact #833: Medieval Knights Were Obsessed with Jousting

Jousting tournaments were so popular in medieval times that knights would often train for months. It was the ultimate reality TV of the Middle Ages, complete with drama, rivalry, and occasional injuries.

Fact #834: A Man Once Ran for President in a Raccoon Hat

In 1840, William Henry Harrison campaigned for U.S. president wearing a raccoon hat. His slogan was "Tippecanoe and Tyler Too," proving political branding has always been a thing.

Fact #835: The Leaning Tower of Pisa Was Never Straight

The Leaning Tower of Pisa started tilting during construction in the 12th century due to unstable soil. Instead of fixing it, they just kept building—talk about a "go with the flow" attitude.

Fact #836: George Washington Grew Cannabis

Washington grew hemp at Mount Vernon, not for recreational use but for making rope and fabric. Even the first president knew the value of a versatile plant.

Fact #837: Victorian Doctors Prescribed Heroin

In the 19th century, heroin was marketed as a cough suppressant and pain reliever. It wasn't until later that people realized, "Hey, this might be a problem."

Fact #838: People Once Feared Coffee

In the 1600s, some people believed coffee was "Satan's drink" because it was so energizing. Even Pope Clement VIII had to step in and declare it harmless. Good thing, or mornings would be tragic.

Fact #839: The Oldest "Your Mom" Joke is Over 3,500 Years Old

An ancient Babylonian tablet contains what is believed to be the world's oldest "your mom" joke. The humor survives, but unfortunately, the punchline has been lost.

Fact #840: The Eiffel Tower Shrinks in Winter

The Eiffel Tower is made of iron, which contracts in cold temperatures, causing it to shrink by about six inches every winter. Even iconic landmarks need a break.

Fact #841: A Medieval Tournament Ended With Cheese

During a 15th-century jousting match in France, the winner's prize was not gold but a wheel of cheese. Because even medieval knights knew cheese was worth fighting for.

Fact #842: President Taft Got Stuck in a Bathtub

William Howard Taft, the 27th U.S. president, once got stuck in a White House bathtub. The solution? They ordered a custom tub big enough to fit four men. Problem solved.

Fact #843: Ancient Romans Invented Concrete

The Romans created concrete that has survived for over 2,000 years. Modern concrete often doesn't last a century—turns out the Romans were way ahead of their time.

Fact #844: The First Alarm Clock Could Only Ring at One Time

The first mechanical alarm clock, invented in 1787, could only ring at 4 a.m. Early risers loved it; everyone else probably hated it.

Fact #845: A Dog Became Mayor of a Town

In 1981, a dog named Bosco became the mayor of Sunol, California. He "served" for 13 years, proving that sometimes a wagging tail is all the leadership you need.

Fact #846: Pirates Wore Eye Patches for Night Vision

Pirates weren't always missing an eye—some wore patches to keep one eye adjusted to the dark. It made it easier to see below deck or during night raids. Smart and sneaky!

Fact #847: King Louis XIV Took Only Three Baths in His Life

The "Sun King" of France reportedly bathed only three times in his entire life. Hygiene wasn't his thing, but wigs and fashion definitely were.

Fact #848: The Mona Lisa Has Her Own Mailbox

At the Louvre in Paris, the Mona Lisa has her own mailbox for fan letters. Yes, even a 500-year-old painting gets more mail than most of us.

Fact #849: A Man Lived Inside a Whale—Sort Of

In 1891, James Bartley claimed he was swallowed by a whale and survived. While the story is likely a myth, it earned him the nickname "The Modern Jonah."

Fact #850: Abraham Lincoln Was a Wrestling Champion

Before becoming president, Lincoln was a skilled wrestler. He reportedly won nearly 300 matches and only lost one. Imagine a Lincoln vs. T. Rex showdown—our money's on Abe.

Fact #851: A City Fought Over a Sausage Recipe

In 1510, two German towns went to court over who invented bratwurst. The debate got so heated that they wrote down the recipe to settle the matter. Priorities, right?

Fact #852: The White House Used to Be a Party Spot

In the early 1800s, anyone could show up to the White House on Inauguration Day. Andrew Jackson's party got so rowdy that he had to escape out a window.

Fact #853: People Used Pineapples as Status Symbols

In the 1700s, pineapples were so rare and expensive in Europe that people would rent them to display at parties. Forget eating them—they were just for showing off.

Fact #854: A Town Once Printed Money on Wood

During World War I, a German town ran out of paper for money and started printing banknotes on wood. It worked so well that people began collecting them as souvenirs.

Fact #855: Vikings Used Animal Bones for Ice Skates

To glide across frozen lakes, Vikings strapped animal bones to their feet. No fancy skates, just raw Viking ingenuity.

Fact #856: The World's Longest Hiccup Attack Lasted 68 Years

Charles Osborne started hiccupping in 1922 and didn't stop until 1990. That's over 430 million hiccups—and probably a lifetime supply of frustration.

Fact #857: Peter the Great Imposed a Beard Tax

In 1698, Russian Tsar Peter the Great wanted to modernize his country, so he taxed men who wore beards. If you wanted to keep your facial hair, you had to pay up—and carry a "beard token" as proof.

Fact #858: Einstein's Last Words Are a Mystery

Albert Einstein's final words were spoken in German, but the nurse attending him didn't speak the language. Whatever he said, it's lost to history forever.

Fact #859: A Tree Owns Itself

In Athens, Georgia, there's a tree that legally owns itself. The land it stands on was deeded to it in the early 1800s. Talk about rooting for independence.

Fact #860: The Eiffel Tower Was Almost a Giant Guillotine

Before becoming the iconic tower we know today, one proposed design for the Eiffel Tower included a massive guillotine to symbolize the French Revolution. Thankfully, they went with the less gruesome version.

Fact #861: Imelda Marcos Built a Lavish Party Palace

Imelda Marcos constructed the Coconut Palace in Manila, a luxurious mansion made from coconut wood and shells. Originally intended to host Pope John Paul II, he declined the invitation, deeming it inappropriate amidst widespread poverty.

Fact #862: An Emu War Happened in Australia

In 1932, Australia declared war on emus—yes, the giant flightless birds—because they were destroying crops. The emus won, proving they're more than just oversized chickens.

Fact #863: A Stolen Napoleon Statue Became a Toilet

After Napoleon's defeat, a statue of him was melted down and turned into public toilets in England. A humbling end for the emperor, to say the least.

Fact #864: A Dead Pope Was Put on Trial

In 897, Pope Formosus's corpse was dug up, dressed in papal robes, and put on trial for alleged crimes. The bizarre event, known as the Cadaver Synod, ended with his body being thrown into the Tiber River.

Fact #865: Thomas Edison Electrocuted an Elephant

In 1903, Thomas Edison electrocuted an elephant named Topsy to demonstrate the dangers of alternating current (AC). It was a dark PR stunt in his feud with Nikola Tesla.

Fact #866: A Woman Survived a Fall From 33,000 Feet

In 1972, flight attendant Vesna Vulović survived a 33,000-foot fall after her plane exploded mid-air. She holds the Guinness World Record for surviving the highest fall without a parachute.

Fact #867: A Man Invented Lobster Claw Bands to Save Fingers

In the 1800s, fishermen kept losing fingers to angry lobsters. The invention of rubber bands for lobster claws saved countless digits—and made lobster dinners much safer.

Fact #868: The Statue of Liberty Was Almost Gold

When designing the Statue of Liberty, artist Frédéric Auguste Bartholdi considered coating it in gold. Luckily, they went with copper instead—imagine the glare on a sunny day!

Fact #869: There's a Secret Apartment in the Eiffel Tower

Gustave Eiffel, the tower's designer, built a private apartment at the top of the Eiffel Tower where he entertained guests. It's still there today, but sadly, it's not an Airbnb.

Fact #870: An Entire Town Disappeared in Canada

In 1930, the residents of Angikuni Lake, a remote Canadian village, vanished without a trace. To this day, no one knows what happened to them, making it one of Canada's greatest mysteries.

Fact #871: Pigeons Were War Heroes

During World War I and II, pigeons were used to carry messages. One bird, Cher Ami, delivered a message that saved nearly 200 soldiers despite being shot. Pigeon of the year, every year.

Fact #872: A Man Survived Seven Lightning Strikes

Roy Sullivan, a park ranger, survived being struck by lightning seven times between 1942 and 1977. He holds the world record for the most lightning strikes survived—though it's not exactly an award anyone wants.

Fact #873: A President Had Pet Alligators

John Quincy Adams, the 6th U.S. president, kept pet alligators in the White House. He housed them in the East Room, where they reportedly scared guests. Practical joke or pet love? You decide.

Fact #874: People Once Paid to Watch Grass Grow

In 19th-century England, wealthy landowners hosted "grass-watching parties" where guests literally sat and watched the grass grow. Talk about entertainment before Netflix.

Fact #875: A Typo Caused a Shipwreck

In 1914, a lighthouse keeper in Scotland accidentally typed the wrong coordinates in a message, causing a ship to crash. That's why proofreading is important, even for lighthouses.

Fact #876: The Moon Smells Like Gunpowder

Astronauts from the Apollo missions reported that lunar dust smelled like gunpowder. So, the Moon isn't just out of this world—it smells explosive too.

Fact #877: A Parrot Testified in Court

In 1993, a parrot in India became a witness in a murder case after it repeatedly said the victim's name in distress. It's not clear if Polly got crackers for its service.

Fact #878: The First Vending Machine Sold Holy Water

The first vending machine, invented in ancient Greece, dispensed holy water. You'd insert a coin, and the machine would release a small amount. A divine invention for its time.

Fact #879: The Eiffel Tower Was Once a Billboard

In the early 20th century, the Eiffel Tower was used as a massive billboard for Citroën cars. It lit up with the brand's name, making it the most stylish ad of its time.

Fact #880: The Great Emu War Inspired a Video Game

The infamous Australian Emu War of 1932 inspired a comedic video game where players control emus fighting against humans. Gaming history gets weird, too.

Fact #881: Medieval Helmets Had Straw Straws

Knights sometimes used helmets with built-in straws to drink water or ale during battles. Fighting was tough work, and hydration was key—even if it came through a medieval sippy cup.

Fact #882: A Monkey Was Once Tried in Court

In 1825, a monkey was put on trial in Hartlepool, England, after locals thought it was a French spy. The monkey didn't fare well, but the town now celebrates it with a festival.

Fact #883: Ben Franklin Wanted the Turkey as the U.S. Symbol

Benjamin Franklin thought the bald eagle was "a bird of bad moral character" and suggested the turkey instead. He described the turkey as a "much more respectable bird," but let's be honest—it would have made for a much less intimidating national emblem.

Fact #884: The Oldest Known Joke is a Fart Joke

The world's oldest recorded joke dates back to 1900 BCE in ancient Sumer and involves flatulence. Proof that humans have always found bodily functions hilarious.

Fact #885: Shakespeare Invented Words We Still Use Today

William Shakespeare coined over 1,700 words, including "eyeball," "swagger," and "bedazzled." Without him, modern slang would be a lot less fun.

Fact #886: A Rooster Was Once Convicted of Witchcraft

In the 1600s, a Swiss court put a rooster on trial for laying an egg, which was considered a sign of witchcraft. The poor bird didn't win its case.

Fact #887: The Eiffel Tower Grows in Summer

Thanks to thermal expansion, the iron in the Eiffel Tower expands in the heat, causing it to grow by about six inches in summer. Even landmarks get a little taller in warm weather.

Fact #888: A Pig Was Knighted in Denmark

In the 15th century, a pig was knighted in Denmark for saving a child from a wolf attack. Sir Oinkington became a local hero—and the best-dressed pig in town.

Fact #889: A Man Won the Lottery Twice Using the Same Numbers

In 1984, a man in Virginia won the lottery twice in one day by using the same set of numbers. His odds of winning once were slim—winning twice was practically impossible.

Fact #890: The Library of Congress Has a Vault of Smells

The Library of Congress preserves historic smells, including the scent of old books, leather bindings, and even historical perfumes. It's like a museum for your nose.

Fact #891: The Longest Time Between Twins Being Born is 87 Days

In 2012, a woman gave birth to her first twin prematurely, and the second was born 87 days later. Both babies survived and became a medical marvel.

Fact #892: A City in Canada Has a UFO Landing Pad

St. Paul, Alberta, built the world's first UFO landing pad in 1967. It's still waiting for its first extraterrestrial visitor—but they're prepared just in case.

Fact #893: A Traffic Jam Lasted 12 Days

In 2010, a traffic jam in China stretched over 60 miles and lasted 12 days. Drivers reportedly spent their time playing cards and selling snacks. Road rage was probably off the charts.

Fact #894: A Woman Wrote a Cookbook About Cooking for Aliens

In 1982, a woman published *Intergalactic Cuisine*, a cookbook filled with recipes designed for aliens. It's unclear if any extraterrestrial foodies have tried her dishes yet.

Fact #895: The First Webcam Was Invented to Watch Coffee

The very first webcam was created at Cambridge University to monitor a coffee pot. Scientists didn't want to waste trips to an empty pot—talk about priorities!

Fact #896: An Octopus Once Escaped from an Aquarium

An octopus named Inky escaped from a New Zealand aquarium by squeezing through a small drainpipe that led to the ocean. If octopuses ever form a union, Inky will be their leader.

Fact #897: A Dog Helped Discover Penicillin

Alexander Fleming, who discovered penicillin, was inspired by his dog, who accidentally knocked over a petri dish. The resulting mold sparked one of the greatest medical breakthroughs in history.

Fact #898: A City in Poland Had a Bear for a Soldier

During World War II, a bear named Wojtek served in the Polish army, carrying heavy supplies and even helping in battles. He was promoted to corporal and retired as a national hero.

Fact #899: The First Olympic Marathon Had a Cheater

In the 1904 Olympics, a runner named Fred Lorz hitched a ride in a car for part of the marathon. He later claimed it was a joke, but officials weren't laughing.

Fact #900: A President Gave a Two-Hour Speech and Got Sick

In 1841, William Henry Harrison delivered the longest inaugural address in U.S. history, lasting nearly two hours, on a freezing, rainy day without a coat or hat. Shortly after, he caught a cold that developed into pneumonia and died 31 days into his presidency. It's a cautionary tale about knowing when to keep your speeches short and your coat on.

POP QUIZ

Think you're ready to take on the quirkiest moments in history? Let's see how much you remember from Chapter 9's wild ride through the past!

Questions

1. **True or False:** Napoleon Bonaparte was once attacked by rabbits.
2. Cleopatra lived closer to which event: the building of the Great Pyramid of Giza or the Moon landing?
3. **True or False:** A chicken survived for 18 months after losing its head.
4. What unusual tax did Peter the Great impose in Russia?
5. **True or False:** The Eiffel Tower shrinks in the winter.
6. What did Benjamin Franklin suggest as the national symbol of the United States instead of the bald eagle?
7. **True or False:** The world's first vending machine dispensed coffee.
8. Which animal "served" as a soldier in the Polish army during World War II?
9. **True or False:** Albert Einstein's brain was stolen after his death.
10. What led to the U.S. and Britain almost going to war in 1859, an event known as the "Pig War"?

Answers

1. **True:** Napoleon was attacked by rabbits during a hunting trip. Fluffy rebellion at its finest!
2. **Moon Landing:** Cleopatra lived closer to Neil Armstrong than to the Great Pyramid's construction.
3. **True:** Mike the headless chicken lived for 18 months thanks to careful feeding.
4. **Beard Tax:** Peter the Great taxed beards to modernize Russia with clean-shaven faces.
5. **True:** The Eiffel Tower shrinks six inches in winter due to thermal contraction.
6. **Turkey:** Franklin thought turkeys were nobler than bald eagles. Imagine them on U.S. currency!
7. **False:** Ancient Greece's first vending machine dispensed holy water, not coffee.
8. **Bear:** Wojtek the bear served in WWII, carrying supplies and earning the rank of corporal.
9. **True:** After his death, Einstein's brain was taken without consent and studied by scientists for decades.
10. **Pig War:** A farmer shooting a neighbor's pig sparked a standoff, but thankfully no humans were harmed.

How did you do? Whether you aced it or learned something new, history's quirks always keep us on our toes!

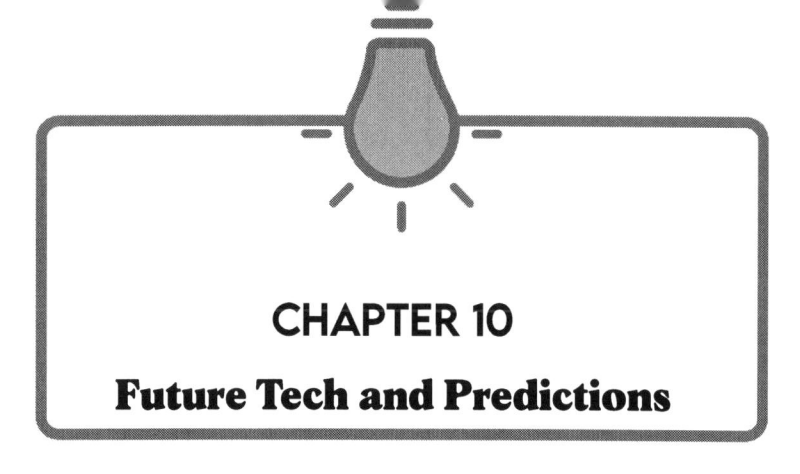

CHAPTER 10

Future Tech and Predictions

The future is closer than you think! From mind-blowing technology to strange innovations and daring predictions, this chapter dives into the wild possibilities that await humanity. Whether it's robots cooking your dinner or humans vacationing on Mars, the future promises to be anything but boring.

Fact #901: AI Can Write Music (and It's Not Half Bad)

Artificial Intelligence can compose music, and some of it is surprisingly good. One day, your favorite pop star might just be a robot—but at least they won't miss any notes during a concert.

Fact #902: Self-Driving Cars Can Have Road Rage (Sort Of)

Self-driving cars are programmed to be polite, but when they encounter aggressive human drivers, they often overcompensate by being too cautious. Even robots can get flustered in traffic.

Fact #903: 3D Printing Can Create Food

Fancy a pizza printed on demand? 3D printers can now make food layer by layer. The only downside? You can't print a refund if the pizza tastes bad.

Fact #904: Robots Are Learning to Have a Sense of Humor

Some AI robots are being trained to understand humor. While they're still figuring out dad jokes, they might soon be cracking one-liners better than your funniest friend.

Fact #905: Humans Might Live on Floating Cities

With rising sea levels, scientists are designing floating cities that could house thousands of people. Think cruise ships, but way bigger—and hopefully with less seasickness.

Fact #906: Space Tourism Could Be the Next Big Thing

Companies like SpaceX and Blue Origin are already offering spaceflights to paying customers. The view from your hotel window might soon include Earth from orbit. Just don't forget your zero-gravity slippers.

Fact #907: Holographic Calls Are Coming

Forget boring video calls. Holograms will let you project a 3D version of yourself anywhere. Your next family Zoom call might look like a scene from *Star Wars*.

Fact #908: Lab-Grown Meat is a Real Thing

Scientists can now grow meat in a lab without ever needing a farm. It's great for reducing animal farming, but imagine explaining your "petri dish burger" at a dinner party.

Fact #909: Wearable Tech Will Know Your Mood

Future wearable devices might track your emotions, helping you manage stress or boost happiness. They'll probably also remind you when you've had too much coffee.

Fact #910: Humans Are Growing Organs in Labs

Doctors are growing organs like hearts and kidneys in labs, potentially eliminating transplant waiting lists. It's like ordering a replacement part for your body—but way cooler.

Fact #911: Virtual Reality Will Make Vacations Cheaper

Can't afford a trip to Hawaii? Virtual Reality will let you "visit" beaches, cities, and even other planets without leaving your living room. Sand not included.

Fact #912: Delivery Drones Are Taking Over

Forget delivery drivers—drones are already being tested to drop packages right at your doorstep. No tips required, but they might steal your food.

Fact #913: Smart Clothes Are the Future

Clothes embedded with tech could monitor your health, adjust to the weather, or even charge your phone. Just hope they don't get a software update while you're wearing them.

Fact #914: Flying Cars Are Actually Happening

It's not just sci-fi anymore—flying cars are being developed for urban transportation. The bad news? You'll still probably get stuck in sky traffic.

Fact #915: Robots Are Becoming Chefs

In the future, robots might cook your meals. Some restaurants already use robot chefs to prepare dishes, but good luck getting them to customize your order.

Fact #916: Humans Could Be Cryogenically Frozen

Cryonics technology aims to freeze humans until future medicine can cure their diseases. It's like hitting the "pause" button on life—minus the remote.

Fact #917: Personalized Medicine Will Be Tailored to Your DNA

Doctors will use your genetic code to create treatments customized just for you. In the future, "take two pills and call me in the morning" might be a thing of the past.

Fact #918: Robots Could Become Your Best Friends

Social robots like Pepper and Buddy are designed to keep people company. They'll never flake on plans, but don't expect them to laugh at all your jokes.

Fact #919: Colonies on Mars Could Happen Soon

Space agencies and private companies are already planning how to build habitats on Mars. Your grandkids might grow up with red dirt under their feet—and a Martian mailing address.

Fact #920: Brain Implants Could Make You Smarter

Neural implants, like Elon Musk's Neuralink, aim to enhance human intelligence. Someday, you might download knowledge directly into your brain—no studying required.

Fact #921: AI Could Predict Your Future

Advanced AI systems are being developed to predict health risks, weather patterns, and even personal behavior. It's like having a fortune teller, but with way more data.

Fact #922: Robots Can Now Paint Masterpieces

AI artists like DALL-E can create stunning artwork in seconds. The art world is divided—are they creative geniuses or just good at copying styles?

Fact #923: The Moon Could Have Its Own Internet

NASA is planning to set up Wi-Fi on the Moon to help astronauts communicate. Moon selfies are about to become the ultimate social media flex.

Fact #924: Smart Glasses Will Be the Next Smartphone

Augmented reality glasses will let you browse the web, take photos, and check your messages—all while looking like you're just staring into space. Welcome to the future of multitasking.

Fact #925: Robots Could Win Nobel Prizes

With AI making breakthroughs in science and medicine, it's only a matter of time before a robot takes home a Nobel Prize. Hopefully, they won't write their acceptance speech in binary.

Fact #926: The Sun Could Power the Entire Earth

Scientists are developing solar panels so efficient that just a small percentage of sunlight could power the whole planet. Solar power might soon be the ultimate energy source.

Fact #927: AI Could Write Bestselling Novels

AI is already writing books and screenplays. Someday, your favorite author might just be a clever algorithm—though hopefully, it learns to avoid clichés.

Fact #928: Scientists Are Developing Invisibility Cloaks

Yes, Harry Potter fans, invisibility cloaks are real—at least in prototype form. Researchers are creating materials that bend light, making objects disappear. Hide-and-seek will never be the same.

Fact #929: Mind-Controlled Tech is Coming

Devices controlled by your brainwaves are already being tested, from wheelchairs to drones. The next frontier? Mind-controlled video games. Game on!

Fact #930: Humans Might Live to Be 150

With advances in medicine and anti-aging research, some scientists believe humans could live up to 150 years. That's a lot of birthday candles.

Fact #931: People Are Growing Vegetables in Space

Astronauts have successfully grown lettuce, radishes, and even chili peppers on the International Space Station. Space salads are officially on the menu.

Fact #932: AI is Writing Movie Scripts

Some movies and short films have already been scripted by AI. They're not Oscar-worthy yet, but they're proof that the robots might steal Hollywood's spotlight someday.

Fact #933: Virtual Reality Could Replace Work Meetings

Tired of Zoom? VR headsets will soon let you meet with coworkers in a virtual office. Just don't forget to dress your avatar professionally—pajamas aren't business casual.

Fact #934: Smart Toilets Can Monitor Your Health

Future toilets could analyze your waste for early signs of disease. Imagine getting a health report from your toilet every morning. Gross but useful.

Fact #935: The Oceans Could Power the World

Scientists are working on ways to harness wave and tidal energy to generate electricity. The ocean might soon be your home's power source—just don't forget your surfboard.

Fact #936: Robots Can Be Emotional Support Companions

Robots like Paro the Seal are designed to comfort people, especially in hospitals or nursing homes. They're cute, calming, and they never shed.

Fact #937: Electric Planes Are in Development

The aviation industry is exploring electric planes that could make air travel quieter, greener, and cheaper. Your future flights might run on batteries, not jet fuel.

Fact #938: Space Elevators Might Become a Reality

Scientists are working on building a space elevator—a giant cable connecting Earth to space. It sounds like science fiction, but it could revolutionize space travel.

Fact #939: AI Can Diagnose Diseases Faster Than Doctors

AI systems are being trained to detect diseases like cancer and heart conditions earlier and more accurately than humans. The doctor of the future might just be an algorithm.

Fact #940: Mars Might Have Underground Cities

If humans colonize Mars, we'll likely build underground to protect against radiation and dust storms. Living like mole people on the Red Planet? Sign us up.

Fact #941: AI Can Predict Natural Disasters

Advanced AI programs can analyze patterns to predict hurricanes, earthquakes, and floods. It's like having a digital fortune teller for the weather.

Fact #942: Smart Contact Lenses Will Do More Than Help You See

Future contact lenses will have built-in displays, allowing you to check emails, navigate maps, or even monitor your health—all with just a blink.

Fact #943: Earthquakes Could Be Controlled

Some scientists believe we could use technology to release tectonic pressure gradually, preventing devastating earthquakes. Mother Nature might finally get a little human assistance.

Fact #944: Humans Could Hibernate for Space Travel

Scientists are studying ways to put astronauts in a hibernation state for long space journeys. Think *Sleeping Beauty*, but on a spaceship.

Fact #945: Nanobots Could Swim Through Your Bloodstream

Tiny robots smaller than a grain of sand could one day be injected into your body to target diseases, repair cells, or deliver medicine. Microscopic healers are the future.

Fact #946: Humans Might Grow Extra Limbs with Biotechnology

With advances in genetics and biotechnology, some scientists believe humans could eventually grow extra limbs or organs. Imagine having an extra arm for multitasking!

Fact #947: The Moon Could Have Cities by 2050

With NASA and private companies planning lunar missions, experts predict the Moon could have permanent human settlements within a few decades. Moon pizza delivery, anyone?

Fact #948: AI Will Learn to Write Jokes

Some AI systems are already trying their hand at comedy. While they're not exactly ready for stand-up, one day they might actually make you laugh.

Fact #949: Virtual Pets Are Becoming Surprisingly Realistic

Future virtual pets will be so advanced that you'll feel like they're alive. They'll wag their tails, respond to commands, and even snuggle—no cleanup required.

Fact #950: Smart Cities Will Think for Themselves

Smart cities of the future will use AI to manage traffic, reduce energy waste, and improve public safety. It's like having a city that's smarter than most of its residents.

Fact #951: Plastic-Eating Enzymes Could Save the Planet

Scientists have created enzymes that break down plastic in hours instead of centuries. This game-changing tech could make a big dent in global pollution.

Fact #952: Space Junk Removal Is Becoming a Priority

With so much debris floating in orbit, companies are designing "space janitors" to clean it up. The future of space travel might depend on a good cleanup crew.

Fact #953: Smart Tattoos Will Monitor Your Health

Temporary tattoos made with special ink will one day track your heart rate, hydration levels, and more. They're part tattoo, part wearable tech, and fully awesome.

Fact #954: AI Could Compose Entire Symphony Orchestras

AI programs like OpenAI's MuseNet can already compose symphonies. The future of classical music might come from a computer, not a composer.

Fact #955: Underwater Cities Could Be a Thing

As land becomes scarce, some scientists envision entire cities built under the ocean. They'd come with stunning views of marine life—just watch out for sharks.

Fact #956: Robots Could Teach Classrooms

AI-powered robots might one day replace human teachers. They'll never lose patience, but don't expect them to grade on a curve.

Fact #957: A Space Ark Could Save Humanity

In the event of a global catastrophe, scientists are working on creating "space arks" to preserve humanity by sending people to other planets. Space Noah, here we come!

Fact #958: AI Could Create Virtual Copies of You

Advanced AI could create digital avatars that think, talk, and act like you. Your virtual twin could handle meetings while you binge-watch TV.

Fact #959: Solar Power Could Be Beamed From Space

Scientists are developing ways to collect solar power in space and beam it to Earth. Future energy grids might be powered from above the clouds.

Fact #960: Artificial Islands Are Being Built for Housing

With land running out, some cities are constructing floating or artificial islands for homes, parks, and businesses. The future of real estate might be on water.

Fact #961: AI Could Replace Weather Forecasters

Advanced AI systems are becoming so good at predicting the weather that they might one day replace meteorologists. Just imagine: no more awkward small talk during your weather updates.

Fact #962: Humans Could Drink Water Made From Thin Air

Scientists are developing machines that extract water from the air, even in arid regions. The future could mean an endless water supply—even in deserts.

Fact #963: Space Mining Could Make Asteroids the New Gold Rush

Companies are planning missions to mine asteroids for precious metals like gold and platinum. In the future, your engagement ring might have cosmic origins.

Fact #964: Robots Could Do Your Grocery Shopping

Robots equipped with AI are already being tested to shop for groceries and deliver them to your doorstep. Forget shopping lists—just let the robots handle it.

Fact #965: AI Could Write Personalized Stories for You

AI systems will soon be able to create custom novels based on your preferences. Whether you want a romance, thriller, or sci-fi, your dream story could be just a click away.

Fact #966: Scientists Are Working on Reversing Aging

Anti-aging research is advancing quickly, with scientists developing treatments to slow or even reverse the aging process. The fountain of youth might be closer than we think.

Fact #967: Future Farms Will Be Vertical and Indoor

Vertical farms, which grow crops indoors using LED lights and hydroponics, are becoming more common. They use less water, less land, and can grow food year-round—even in cities.

Fact #968: Robots Are Learning to Build Robots

Self-replicating robots are no longer science fiction. In the future, robots might build other robots, reducing manufacturing costs and speeding up production.

Fact #969: Wearable Translation Devices Will Break Language Barriers

Devices like earbuds and glasses will soon offer real-time translation, making communication easier than ever. Traveling abroad? No problem—your gadgets will be your personal translator.

Fact #970: Quantum Computers Will Change Everything

Quantum computers, which use the principles of quantum mechanics, will solve problems millions of times faster than today's fastest supercomputers. They could revolutionize medicine, cryptography, and AI.

Fact #971: AI Could Predict Disease Outbreaks

Using data from social media, health records, and climate patterns, AI systems might predict disease outbreaks before they happen, potentially saving millions of lives.

Fact #972: Floating Power Plants Could Provide Energy

Floating wind farms and solar panels are being developed to generate renewable energy at sea. These power plants could help reduce carbon emissions and provide energy to coastal cities.

Fact #973: Smart Fridges Will Cook for You

Future refrigerators might suggest recipes based on the ingredients inside and even preheat your oven. It's like having a chef in your kitchen—minus the chef.

Fact #974: AI Could Design Buildings

Architectural AI programs are already designing buildings that are energy-efficient, earthquake-resistant, and visually stunning. The architects of the future might be algorithms.

Fact #975: Genetic Editing Could Eradicate Diseases

CRISPR technology allows scientists to edit genes, potentially eliminating genetic diseases like sickle cell anemia and Huntington's disease. It's a medical game-changer.

Fact #976: Robots Are Exploring the Deep Ocean

Robotic submarines are venturing into parts of the ocean humans have never seen. They're discovering new species and mapping the seabed—future Jacques Cousteaus with circuits.

Fact #977: The Internet Will Be 100 Times Faster

Advancements in fiber optics and 6G technology could make the internet so fast that downloads will be instant. No more buffering—ever.

Fact #978: AI Could Make Clothes Tailored to You

Imagine ordering clothes that fit perfectly without trying them on. AI will analyze your measurements and preferences to create custom outfits just for you.

Fact #979: Robots Could Help in Natural Disasters

Rescue robots are being developed to search for survivors in earthquakes, floods, and fires. They can go where humans can't, making them lifesaving heroes.

Fact #980: Space Farming is the Next Big Thing

Scientists are experimenting with growing crops on the Moon and Mars. Future astronauts might eat salads grown in space—just add space dressing.

Fact #981: AI Could Teach You New Skills Instantly

AI tutors are being developed to teach people new skills, from playing piano to coding. Personalized lessons mean learning will never be the same.

Fact #982: Solar Roads Could Power Cities

Engineers are designing roads with built-in solar panels to generate electricity. Driving to work could literally help power your office.

Fact #983: Augmented Reality Will Change Shopping Forever

AR will let you try on clothes, see furniture in your home, or preview haircuts—all without leaving your house. Shopping trips will be a thing of the past.

Fact #984: Humans Could Upload Their Minds to Computers

Some scientists believe we'll eventually be able to upload our consciousness to computers, achieving digital immortality. Who needs a body when you can live forever in the cloud?

Fact #985: AI Could Solve the Plastic Problem

Advanced AI systems are being used to identify and sort recyclable plastics more efficiently, helping reduce waste and clean up the planet.

Fact #986: Smart Roads Could Charge Electric Cars

Future roads may have built-in wireless charging for electric cars, letting them recharge as they drive. Long road trips just got a lot easier.

Fact #987: AI Could Help Write Laws

Governments are testing AI systems to help draft legislation, ensuring laws are fair and free from loopholes. Finally, a legal assistant that never takes a coffee break.

Fact #988: Nanotechnology Could Make Clothes Water-Repellent

Nano-coatings will make future clothes resistant to water, stains, and even dirt. Say goodbye to laundry day!

Fact #989: Space Junk Recycling is on the Horizon

Companies are developing ways to recycle space debris into new materials for spacecraft. The ultimate recycling project is happening in orbit.

Fact #990: AI Could Replace Therapists

Virtual therapists powered by AI are being developed to provide mental health support. They're empathetic, always available, and never get tired of listening.

Fact #991: Future Computers Could Be Powered by DNA

DNA-based computers could store far more data than traditional ones and use less energy. The future of tech might be written in genetic code.

Fact #992: Robots Will Farm Underwater

Underwater farming systems are being designed to grow crops like seaweed and shellfish, providing sustainable food sources for the future.

Fact #993: Smart Shoes Will Help You Walk Better

Shoes embedded with sensors and AI will analyze your gait and provide real-time feedback, helping you improve posture and prevent injuries.

Fact #994: AI Could Design Video Games

AI systems are already creating custom levels and characters for games, and in the future, they might design entire games tailored to individual players.

Fact #995: Holograms Could Replace TVs

Holographic displays might soon replace traditional screens, creating immersive viewing experiences. Your next binge-watch could look like it's happening in your living room.

Fact #996: Smart Pill Bottles Will Keep You on Track

Pill bottles with built-in reminders will alert you if you forget to take your medication. No more guessing if you missed a dose.

Fact #997: Electric Buses Will Replace Diesel

Cities worldwide are adopting electric buses to reduce pollution and improve air quality. Future public transport will be greener and quieter.

Fact #998: Floating Libraries Could Travel the World

Designs for floating libraries aim to bring books to coastal communities. Imagine checking out a novel while cruising on the open seas.

Fact #999: AI Could Create Virtual Worlds for You to Explore

AI will build personalized virtual worlds based on your preferences. Whether you want a tropical paradise or a medieval kingdom, your dream escape is just a headset away.

Fact #1000: The Future is Closer Than You Think

From space travel to AI-driven cities, the world of tomorrow is being built today. One thing's for sure—the future will be anything but boring.

POP QUIZ

The future is now—test your knowledge of the weird, wild, and wonderful advancements we're headed toward!

Questions

1. **True or False:** AI can already write music and compose symphonies.
2. What kind of car can charge itself while driving on specially designed smart roads?
3. **True or False:** Scientists have grown vegetables on the International Space Station.
4. What animal inspired the development of underwater farming systems?
5. **True or False:** Future contact lenses might allow you to read emails directly through your eyes.
6. What technology could replace traditional TVs with immersive displays?
7. **True or False:** Astronauts might soon have Wi-Fi on the Moon.
8. What futuristic farming method involves growing crops indoors using vertical space?
9. **True or False:** AI systems are already being tested to diagnose diseases faster than human doctors.
10. What material are future DNA-based computers expected to use for storage?

Answers

1. **True:** AI like MuseNet can compose symphonies, blending tech with Mozart-level creativity.
2. **Electric Cars:** Smart roads with wireless charging will keep EVs powered while driving.
3. **True:** Astronauts have grown lettuce and chili peppers on the ISS. Space salads are here!
4. **Seaweed:** Underwater farms grow seaweed, inspired by eco-friendly marine systems.
5. **True:** Future smart lenses will display emails, maps, and health data right before your eyes.
6. **Holograms:** TVs may be replaced by holograms for immersive viewing experiences.
7. **True:** NASA is planning Moon-based Wi-Fi to improve astronaut communication during missions.
8. **Vertical Farms:** Crops grown indoors in stacked layers save space and resources.
9. AI is being developed to diagnose diseases like cancer, often with promising accuracy.
10. **DNA:** Computers may soon store data using DNA, combining biology with technology.

How did you do? Whether you're an expert in futuristic tech or just learned something new, one thing's for sure—the future is full of surprises!

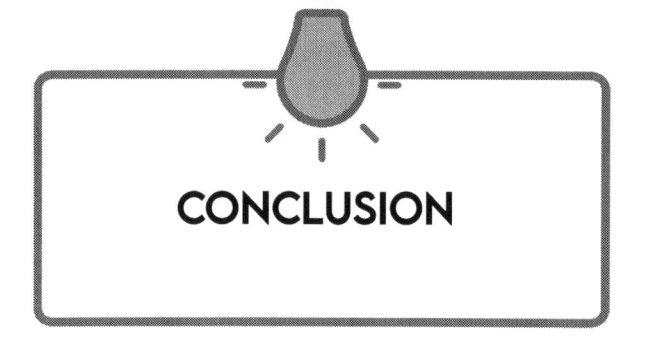

CONCLUSION

And that's a wrap! You've just traveled through 1,000 incredible, funny, and downright bizarre facts that span everything from the mysteries of space to the quirks of history, the wonders of nature, and even the possibilities of the future. Along the way, you've laughed, learned, and maybe even discovered a few new favorite topics to obsess over.

This book was designed to spark curiosity and remind us how fascinating and strange the world (and beyond) truly is. Whether it's the idea of dinosaurs with feathers, a bear serving in the army, or the thought of humans one day living on Mars, the universe never runs out of surprises.

But the best part? This isn't the end. Every day, new discoveries are being made, old mysteries are being solved, and our understanding of the world continues to grow. The facts in this book are just a snapshot of what we know now—who knows what incredible things we'll uncover next?

So, stay curious, keep exploring, and never stop marveling at the amazing world around you. The adventure is far from over—there's always something new to discover!

SHARE YOUR THOUGHTS

Thank you for exploring **The Ultimate Book of Incredible Facts.** If you enjoyed this journey through fun and fascinating facts, it would be wonderful if you could take a moment to leave an honest review.

Simply scan the barcode below to share your experience on Amazon. Your review helps others discover this book and shows your support for creating more exciting collections like this.

Thank you for being part of this adventure!

Printed in Great Britain
by Amazon

53016335R00123